D0414773

The Future of College Mathematics

Proceedings of a Conference/Workshop on the First Two Years of College Mathematics

Edited by
Anthony Ralston
Gail S. Young

Springer-Verlag
New York Heidelberg Berlin

Anthony Ralston
SUNY at Buffalo
Department of Computer Science
4226 Ridge Lea Road
Amherst, N.Y. 14226
U.S.A.

Gail S. Young
University of Wyoming
Department of Mathematics
Laramie, Wyoming 82071
U.S.A.

AMS Subject Classifications: 00A10, 00A25, 00A99, 01A80

Library of Congress Cataloging in Publication Data
Main entry under title:
The Future of College Mathematics
Bibliography: p.
Includes index.
1. Mathematics—Study and teaching (Higher)—
Congresses. I. Ralston, Anthony. II. Young, Gail S.
III. Alfred P. Sloan Foundation.
QA11.A1F84 1983 510'7'11 82-19447

Conference/Workshop held at
Williams College, Williamstown, MA
from June 28 to July 1, 1982.
Sponsored by the
Alfred P. Sloan Foundation.

With 3 Illustrations

Printed and bound by R.R. Donnelley & Sons, Harrisonburg, VA.
Printed in the United States of America.

9 8 7 6 5 4 3 2 1

ISBN 0-387-90813-7 Springer-Verlag New York Heidelberg Berlin
ISBN 3-540-90813-7 Springer-Verlag Berlin Heidelberg New York

Contents

INTRODUCTION

The Conference/Workshop of which these are the proceedings was held from 28 June to 1 July, 1982 at Williams College, Williamstown, MA. The meeting was funded in its entirety by the Alfred P. Sloan Foundation. The conference program and the list of participants follow this introduction.

The purpose of the conference was to discuss the re-structuring of the first two years of college mathematics to provide some balance between the traditional calculus-linear algebra sequence and discrete mathematics. The remainder of this volume contains arguments both for and against such a change and some ideas as to what a new curriculum might look like. A too brief summary of the deliberations at Williams is that, while there were - and are - inevitable differences of opinion on details and nuance, at least the attendees at this conference had no doubt that change in the lower division mathematics curriculum is desirable and is coming.

With only a couple of exceptions, all the papers in this volume were prepared and distributed before the conference. At Williams they were not formally presented but rather discussed under the assumption that all participants had read them. The salient portions of these discussions, containing the essence rather than a verbatim transcription of each discussant's remarks, is included after each paper. Generally these begin with the author's remarks elaborating on his or her paper. The authors had an opportunity to revise their papers in light of the discussions at the conference and, of course, it is these revisions which appear here.

The order of presentation of the papers (see the program following this introduction) reflects much less some inner logic than it does necessity forced by the schedules of some speakers who could not be present for all four days.

The third day of the meeting was devoted to three workshops, two which were concerned with the curriculum itself and the third with the problems of carrying the work of the conference forward to actual implementations of a new curriculum. Following the papers and their discussions are reports from the three workshops.

The workshop reports were presented orally on the final day of the conference. Although there was some general discussion following these reports, it is not summarized in these proceedings. As is often the case in such meetings, after almost three and a half days of talk, there was nothing of significance left to say in the final hour of the conference. I think it is quite proper and appropriate that the reports of the three workshops together, of course, with the papers themselves should stand as the results and conclusions of the conference.

Anthony Ralston

SLOAN CONFERENCE/WORKSHOP

List of Attendees

Richard Alo'	University of Houston
Richard D. Anderson	Louisiana State University
Lida Barrett	Northern Illinois University
Donald Bushaw	Washington State University
Ed Dubinsky	Clarkson College
Stephen Garland	Dartmouth College
Isaac Greber	Case Western Reserve University
John Kemeny	Dartmouth College
Robert M. Kozelka	Williams College
Donald Kreider	Dartmouth College
Jack Lochhead	University of Massachusetts
William F. Lucas	Cornell University
Stephen Maurer	Swarthmore College
Robert Norman	Dartmouth College
Jeffrey Parker	Williams College
H. O. Pollak	Bell Laboratories
Ronald E. Prather	University of Denver
Anthony Ralston	SUNY at Buffalo
Fred S. Roberts	Rutgers University
William Scherlis	Carnegie-Mellon University
Guilford L. Spencer	Williams College
Lynn A. Steen	St. Olaf College
A. W. Tucker	Princeton University
Alan Tucker	SUNY at Stony Brook
Julian Weissglass	University of California, Santa Barbara
Stephen White	Alfred P. Sloan Foundation
Herbert S. Wilf	University of Pennsylvania
Gail S. Young	University of Wyoming
Stanley Zionts	SUNY at Buffalo

PROGRAM

Monday, 28 June 1982

Morning:

1. Opening remarks (Ralston, Young)
2. The Overall Problem - Discussion of Ralston and Young papers
3. Discussion of Wilf and Lucas papers

Evening:

The Effects of a New Curriculum on Other Disciplines - Discussion of Lochhead, Greber, Norman, Zionts and Scherlis-Shaw papers.

Tuesday, 29 June

Morning:

1. Discussion of Steen paper
2. A New Curriculum and Its Possible Formats - Discussion of Bushaw, Roberts, and Tucker papers.
3. Discussion of Alo paper.

Afternoon:

Effects of a New Curriculum in Other Areas - Discussion of Maurer, Weissglass and Albers papers.

Dinner:

Address by John Kemeny

Evening:

Problems of Implementing a New Curriculum - Discussion of Garland-Kreider and Barrett papers.

Wednesday, 30 June

Morning and Afternoon:

Workshops

Thursday, 1 July

Morning:

1. Reports of leaders of three workshops.
2. General discussion and closing remarks.

Note: The paper by R.D. Anderson was written at the meeting itself but was not discussed by the conferees.

AN OVERVIEW OF THE ARGUMENTS CONCERNING THE DEVELOPMENT OF A NEW CURRICULUM

Anthony Ralston
SUNY at Buffalo
Amherst, NY 14226

You don't have to believe that there should be a new curriculum for the first two years of college mathematics to be interested in the results of this conference. You need only believe that the arguments in favor of change are substantial enough to make the exercise of developing a new curriculum worthwhile. For only with a new curriculum available in considerable detail will it be possible to judge with any precision whether it is superior to the current curriculum. And only with a new model will it be possible to refute the arguments against change or, perhaps, to adjudge that these arguments are irrefutable.

From many conversations and much correspondence with mathematicians and others over the past three years I am convinced that

--the current curriculum is not serving as large a proportion of the disciplines that need mathematics as it should

--that mathematics itself is being ill-served by the classical curriculum

--that the mathematics community is much more ripe for change than I had initially believed; it is willing, even perhaps eager to be led

--that the many arguments that can be raised against major curriculum change in mathematics, often well-reasoned and compelling and made by scientists whose stature deserves a hearing, are none of them conclusive and all are, to a degree at least, rebuttable.

But there has been enough of generalities and philosophy. It is time now to show that curriculum change is not just desirable but possible, not merely a general position but one which can be defined in detail, not a program for a distant future but a task on which significant progress can be made this decade.

One of the pitfalls of writing on this subject, a danger I have not entirely avoided, is to treat the problem too simplistically. There is no standard, monolithic curriculum for the first two years of college mathematics in the United States. Too many circumstances in American higher education preclude this:

--the variety of our institutions of higher education

--the variety, probably (unfortunately) the <u>increasing</u> variety of the preparation for the study of mathematics with which students enter college

--the inevitable variety of opinions among professional mathematicians about the nuances, at least, of undergraduate education.

Nevertheless, having paid the obligatory obeisiance to this variety, it is, I think, reasonable for the purposes of argument to act as if there were a standard two-year calculus-linear algebra sequence against which to compare a new model. At least it is true that such a sequence is taught, with relatively minor variations, at many, many colleges and universities. It is no straw person.

<u>The Barriers to Change</u>

In a letter to me two years ago an eminent computer scientist-mathematician warned me against pursuing the course I was on. Don't, he said, "buck the system." To be fair, he was commenting on whether or not one should try to have a separate mathematics curriculum for computer science undergraduates. That was the position from which I began this (quixotic) odyssey (if you'll forgive a mixed classic) but now I firmly believe that what is good for computer science can be good for everyone. Which is to say that I believe in the possibility of a single curriculum to serve all disciplines including mathematics itself.

This is both an intellectual and a pragmatic statement. Its intellectual soundness can be judged in considerable part from the remainder of this volume. Pragmatically, however, it is clear that any attempt to have two separate curricula for the first two years of college mathematics is a non-starter. At small colleges the reasons for this are obvious but even at large universities it would be difficult, indeed, to stir up an enthusiasm to teach yet another special set of courses in addition to those which may now be provided for social science students, management science students, etc.

Nevertheless even if one accepts the desirability of a new two year sequence for everyone, there are a number of practical difficulties in the way of actually effecting the change:

1. Lack of texts and related materials. Since my paper [1] appeared in the Monthly last year, I have had a number of inquiries from people at colleges and universities who would like to teach discrete mathematics at the freshman level but haven't been able to find a suitable text. Can I suggest one? Unhappily I can't. So far as I know there is no suitable text at the present time. I do know a couple that are in the works. Equally important I know that publishers are very interested in such books (see also John Kemeny's paper later in this book). But it is inevitable for some (many?) years that the availability of texts for a new curriculum will be much inferior to the state of affairs for the classical calculus-linear algebra curriculum. This implies that one of the first tasks that must be undertaken if a new curriculum is to have a chance for acceptance is the development of texts; some specific funding for this purpose may be necessary.

2. Problems with service courses. That it will be difficult to persuade other disciplines--particularly in the physical sciences and engineering--that a significant change in the curriculum of the first two years can be effected which will still serve their needs is, I think, clear. To a degree a negative reaction to any proposed change, especially one diminishing the role of calculus, would be reflex. We may believe--I do--that rational argument can persuade most engineers and physical scientists that what will remain will serve their needs adequately. But there will be some, I know, who will claim that no topic in the current curriculum can be dispensed with which will not disastrously damage some advanced course in discipline X. Can such an argument be disarmed? Not frontally, I suspect. Rather, I think, the telling argument will be the one Herb Wilf makes in his paper, namely that "some topics [in the current curriculum] will have to be deleted just to make time available for the persistent newcomers." This may not make our critics happy but they can, perhaps, become convinced that the changes wrought by computers and computer science suggest--no, demand--changes in the basic undergraduate mathematics education of all students which other disciplines will just have to adapt to.

3. Problems with the interface to the high school curriculum. Having now spent a couple of decades building up the Advanced Placement Calculus paraphernalia in the secondary schools, are we going to suggest that it be dismantled because the role of calculus in college mathematics is to be changed? As Steve Maurer argues cogently in his paper, we cannot do so or, at least, we cannot do so quickly. Thus, he argues in favor of the two one-year sequence (one calculus, one discrete mathematics) model rather than the integrated two year sequence. Because how would AP Calculus students fit into the latter? A sound argument and one which needs to be considered carefully. It is an argument more about tactics than strategy. I do not think the idea of an integrated two year sequence should be abandoned because of it. But we must recognize that, whatever changes are recommended for the college curriculum cannot be made in vacuo;

the implications for the high school curriculum must be carefully considered. It is worth noting that the college curriculum in any discipline should inform the high school curriculum in that subject rather than vice versa (i.e., the top-down rather than the bottom-up approach). Thus, we should aim to design the college curriculum as we think it should be, understanding, as we do so, that its implementation in the form designed may have to take place over a period of years as the high school curriculum changes to meet the new demands of the college curriculum.

And, finally but surely not least, there is the inertia of the entire undergraduate educational system. Except for computer scientists, the clientele of service courses provided by mathematics departments are not unhappy (even, maybe, when they should be). True, computer scientists are one of the largest clienteles for service courses from mathematics departments and may soon be _the_ largest but the clamor from them for change has just started to build. As for the mathematics community itself, well, it is as true for it as for any other academic group that it is easiest to teach this year what you taught last year. Substantial curriculum change means work on the part of all members of the mathematics department. Some concrete incentives must be found to overcome this inertia because pure intellectual arguments will not be enough.

Indeed, these barriers and others that can be imagined are real and formidable but not so great (said he, naively) to cause us to lose heart. Maybe they can't be overcome but, if so, this is a theorem that needs a constructive proof.

The Intellectual Argument For and Against Change

Various papers in this volume treat aspects of this question. Here I will only outline briefly what I see to be the main arguments on both sides.

The arguments in favor of changes in the curriculum are all motivated by the changes wrought in the fabric of science, technology and education by computers and computer science.

A short list of these is:

1. Computer science itself has caused mathematicians, scientists and humanists to rethink some of their basic premises on what is important and possible in their disciplines. Some of the flavor of this from the mathematics point of view is contained in Herb Wilf's paper.

2. Computers have allowed new approaches to old subjects which have changed these subjects and, more particularly, changed education about them. This is in part the subject of Bill Lucas's paper. While this conference is not about the use of computers in education (see John Kemeny's paper), it is about how this use in other disciplines requires an understanding of mathematics other than what has traditionally been taught in the first two years of college mathematics.

3. More generally, the second Industrial Revolution which is upon us and which focuses on the immaterial - knowledge, communications, information - in contrast to its predecessor, will force upon all scientists and all educators a new world view which will require an emphasis on discrete mathematics at least equal to that on classical analysis.

But the arguments against major change are not inconsiderable:

1. The current curriculum for the first two years of college mathematics is the result of long development and careful honing. What evidence there is suggests that it has served the mathematics, scientific, engineering and other communities reasonably well and that it continues to do so.

2. More specifically, whatever the merits of the claims on the curriculum that discrete mathematics may legitimately have, these areas of mathematics should not be a major portion of the curriculum of the first two years. In calculus one has not only a centuries old development which should play an important role

in a liberal education but, moreover, this development has resulted in fundamental theorems and general approaches which are just what students should be learning in basic college mathematics. By contrast discrete mathematics has no fundamental theorems and no major unifying ideas to present to the beginning student. Moreover, except for some relatively trivial subject matter, discrete mathematics is harder to grasp than the classical curriculum and, thus, is particularly to be avoided in times when the mathematical preparation of college freshmen appears to be declining, perhaps precipitously.

3. It is a general rule--in mathematics anyhow--that the subject matter of one course is never really understood until it is used in the next course. The current first two years of college mathematics is the place where students begin really to understand high school mathematics through geometric arguments and liberal use of algebra and trigonometry. Even with the use of computer graphics and symbolic systems on computers, the lessons of high school mathematics cannot be reinforced as well with discrete mathematics as with the current curriculum.

These arguments--and others--are considered both in the papers that follow and in the discussions of these papers. If a major curriculum change is to be proposed and is to be successful, it will not only have to be soundly designed with good supporting textual materials but also it will need the solid intellectual foundations which will convince both the mathematics community and the academic community more generally that change is desirable, even necessary.

Is the Time Ripe?

Even if the pragmatic arguments against change do not dismay you and if you are persuaded already by the intellectual arguments for change, it still needs to be asked whether it is propitious now to try and set the juggernaut rolling. After all the arguments favoring more discrete mathematics in the curriculum could have been made before, indeed were in a sense by the publication of "Finite Mathematics" by Kemeny, Snell and Thompson in 1956. Why now? Isn't the failure to achieve widespread acceptance of finite mathematics 25 years ago likely to be repeated now? Will the time not be riper a few years hence?

It is always easier to find reasons why major changes may fail than why they are likely to succeed. This is especially so in college curriculum matters where most change is glacial. It should be, I think, obvious that the changes of the past quarter century mean that historic failure is a poor guide to the chances for present success. The rise of computer science and the proliferation of computers have changed the world--and the world of mathematics--in profound ways. Thus, while the past may provide general guidance on the tactics or strategy of curriculum change, it is not informative on the question of whether the time to try is now.

Would success be more likely three, five, ten years from now? Perhaps but perhaps not, too. There may be a tide in the affairs of mathematicians which must, too, be taken at the flood. There is, I believe, a momentum abuilding for curriculum change in mathematics which needs to be grasped, harnessed, directed now and which may not easily be built again. Somewhat to my surprise, instead of resisting my thoughts on this subject in the three years I have been talking and writing about it, I find the mathematics community--as well as I can judge it from limited but not inconsiderable contacts--ready for change.

And, finally, from the viewpoint of a computer scientist, the change is needed now--or as close to now as can be achieved. Hordes of mathematically illiterate or uneducated computer scientists are already being loosed upon us. You may believe that, although this will result in production of inefficient, undebugged programs, that is merely unfortunate but hardly disastrous. But the systems being built by these computer scientists have the same--or greater--potential for disaster as collapsed roofs of domed stadiums or suspended walkways in hotels. These systems will control nuclear power plants, chemical plants, assembly lines and, more generally, the day-to-day workings of societal systems. Mathematicians as much or more than others should surely understand the need for precision and accuracy in such systems and should, therefore, wish to insinuate the mathematics appropriate to these aims into the curriculum.

Reference

1. Computer Science, Mathematics and the Undergraduate Curricula in Both, *Amer. Math. Monthly*, 88 (1981), 472-485.

DISCUSSION

Ralston: Two questions I've been asked in response to my various writings on the subject of this conference are:

1. What is the relationship to the introduction of discrete mathematics into the curriculum of the first two years and the Mathematical Sciences curriculum proposed by the CUPM Committee chaired by Alan Tucker?
2. What is the connection between this conference and the Sloan Foundation's New Liberal Arts Program?

To the first question my answer is: "Nothing directly". The first two years of the Mathematical Sciences program are quite traditional. However, if discrete mathematics does play a more important role in the first two years, this could, I believe, only serve to strengthen the aims and purposes of the Mathematical Sciences program since that program stressed more modern and discrete areas of mathematics. As to the New Liberal Arts, this conference is related to that program only in the sense that both have been motivated to a considerable degree by the advent of computers and computer science.

I want to make one other point about my own position on this conference to dispel what seems to be a misapprehension on the part of some:

I have never actually proposed and do not believe that the traditional calculus - linear algebra curriculum should be displaced by discrete mathematics. I believe only that balance between the two is desirable.

Bushaw: The problem of lack of textual materials for a new lower division curriculum is a serious one. One of the most constructive things this conference could do would be to provide guidance to anyone who wishes to do serious writing on this subject.

Roberts: I would hope we would develop here a reasonably detailed description, at least in terms of chapter headings, of what we think is involved in discrete mathematics to be taught at the freshman and sophomore levels.

Wilf: Our major task is to figure out what parts of the current curriculum can safely go, what parts of computer-related mathematics should come in and, most difficult of all, how to marry these two together.

Pollak: What about the size of the "chunks" we have to deal with? Can we consider doing a chapter of this, a chapter of something else and putting them together in a pattern rather than trying for whole courses?

Barrett: That idea is not so very foreign from what is happening because of the need for textbooks to accommodate to both quarter and semester systems.

Lucas: I don't care so much how you package a new curriculum. The larger issue is whether you can create an integrated curriculum with a good deal of integrity like the present calculus courses. Current finite mathematics courses at the lower division level usually do not have this.

Kreider: Suppose we consider the question: If calculus suddenly went away, where and in what portions would it naturally reappear because of the needs of other disciplines? And then, if we take those portions that would reappear, would we still be able to have an integrated, coherent course?

Lochhead: If a movement like this is to succeed, we must pay proper attention to fundamental thinking processes. If we do, I think we will find that these are similar regardless of the particular areas of mathematics in the curriculum of the first two years.

Tucker (Alan): I am probably more skeptical about the need for change in the curriculum than most of you. The argument for more finite mathematics for computer science students in the freshman and sophomore years is unassailable. But maybe a first step should be to build on this natural linkage rather than spreading a new curriculum broadly. On the other hand, I note how watered down freshman calculus has become. Perhaps it should become an honorable sophomore course with the standard freshman course being an honorable finite mathematics course with links to the old college algebra courses.

Anderson: Three points relevant to this discussion which were not emphasized in the papers are:

1. The use of calculus as an intellectual screen
2. Calculus courses are where students learn many algebraic techniques
3. Calculus provides a natural place for the development of geometric intuition

Any changes made must keep these values of the calculus sequence.

Young: We must remember here that, like it or not, there will be a revolution in mathematics education in the next 10-15 years brought on by the use of calculators in the schools and symbolic systems, such as Herb Wilf talks about in his paper, in the colleges.

Maurer: On the issue of coherence, I think it's important that we recognize, as Bill Lucas did, the lack of coherence in current finite mathematics courses. Now, although finite mathematics doesn't have a fundamental theorem like calculus, it does have fundamental ideas like induction and recursion around which a coherent approach can be built.

WHO TAKES ELEMENTARY MATHEMATICS COURSES? WHY? A GUESS, AND SOME PROBLEMS FOR CHANGE

Gail S. Young, University of Wyoming

The sort of statement I would like to be able to fill this paper with is this: "There are presently 72,500 undergraduates taking calculus because they know they need that mathematical background for graduate work for an MBA." It would be even more satisfying to add: "If a course in mathematics were planned specifically for the needs of the MBA, then 173,400 students each year would elect it."

Unfortunately, there do not appear to be sources of data to permit making statements like the first. Given a large grant, and a couple of years, one could get such figures.

For the second statement-- Even if on our second day the conferees had taken a break and outlined the ideal course, even if it were then enthusiastically endorsed by every graduate school of management, no such numerical statement could be made. In the '60s, the CUPM Teacher Training Panel recommended that every prospective elementary school teacher take two years of mathematics specially intended for his needs. We were endorsed by the national organization of state directors of teacher certification, by the National Council of Teachers of Mathematics, by AAAS, etc. We spent several hundreds of thousands of dollars on propagation, and several years of hard work. At the end of the time, the number of colleges requiring two years had nearly doubled, from four to seven. The overwhelming logic of a position does not necessarily lead to action.

Nevertheless some quantitative data are available. The first ones I will cite come from the annual studies of entering freshmen made by the American Council of Education. I have here my rearrangement of a summary of such studies made by the CBMS Survey Committee in Vol. VI (1981) of their continuing reports [1]. These Surveys are the most useful data source for studies like ours. I hope rearrangements of CBMS will permit their continuance.

Table 1. Proposed major fields of entering freshmen, in percent of all freshmen

Subject Area	1966	1970	1975	1980
Humanities and arts	24.3	31.1	12.8	8.9
Social sciences	8.2	8.9	6.2	6.7
Education	10.6	11.6	9.9	7.7
Business	14.3	16.2	18.9	23.9
Biological sciences	10.9	12.9	17.5	17.8
Physical sciences	3.3	2.3	2.7	2.0
Engineering	9.8	8.6	7.9	11.8
Math/stat	4.5	3.2	1.1	.6
Other technical	2.2	3.7	8.6	8.1
Undecided/other	11.8	11.6	14.5	12.4
Total number full-time freshmen	1,163k	1,617k	1,761k	1,712k

The classification has its problems. The first problem one notices is the absence of computer science. Unfortunately ACE lumped that into "Other technical." In 1980, a separate figure was given of 4.9% for computer science, data processing, and computer programming. We all know, however, how fast computer science enrollments are growing.

We also all know in how many cases the degree plans of the freshman are changed or are changed for him. There is also often student confusion; the freshman may have a clear and correct idea of what he will do, but just not know the right name for it.

We can, in fact, tell what the freshmen ended up in by looking at the degree figures in Table 2, from the Department of Education [2]. Unfortunately, the classification used in Table 2 does not coincide with the ACE classification used in Table 1. To all of the classifications in Table 2 should be added "and related fields." Natural Sciences includes, for example, agriculture and nursing. I don't know what happened to "Education." The secondary teachers, I suspect, are listed in their teaching areas, and the elementary teachers under social science. I point out, also, that it is not useful to compare column with column, since Table 2 deals with graduating seniors and Table 1 with freshmen.

Table 2. Bachelor's degrees by area, in percentages of total

Subject Area	1960-61	65-66	70-71	75-76	79-80
Humanities	14.5	17.3	17.3	16.1	14.7
Social sciences	38.0	44.9	47.2	42.5	36.7
Business & management	15.6	12.7	14.3	16.5	19.8
Natural sciences	31.8	25.0	21.2	24.9	28.9
Biological sciences	4.5	5.4	4.4	6.2	6.3
Physical sciences	4.2	3.8	2.6	2.4	2.7
Engineering	10.1	7.6	6.2	5.3	8.4
Math/stat	3.6	4.0	3.1	1.8	1.0
Computer science	0	0	.2	.7	.9
EMP* Total	17.9	15.4	12.1	10.2	13.0
Total bachelor degrees	358k	503k	810k	868k	879k

*EMP = Engineering, Mathematical sciences (including computer science), Physical sciences

In 1975, for example, 18.9% of all freshmen planned degrees in business, and in 1979-80, 19.8% of the bachelor's degrees were, in fact, in business. That apparent fixity of purpose could mask considerable change, but I suspect that of all the classifications, this had the least. In my observation, the typical BBA student is in the first generation of his family to enter college, comes from a lower-middle class background, and looks on college strictly as a vocational school. Since the work is, in most business schools, not all that hard, he stays put.

The social sciences, however, clearly have a different pattern. Only 6.2% of the 1975 freshmen planned degrees in social science; in 1980, 36.7% of the degrees were in social science. We see now one of the practical problems of implementation of recommendations we might make. The decision to switch into social science is made, usually, rather late. Consider several possible changes. We take an English major who decides late in his sophomore year to switch to a social science. If he has taken any mathematics in college at all, it has been a course required for all students (who did not take something more advanced). He will now resist as hard as he can taking a difficult freshman mathematics course. Or consider a pre-med who has met organic chemistry and abandoned his plans. He has had a course in calculus, and a lot of science, but is behind on courses associated with his new major. He will also, because of this time pressure, resist more mathematics. For a third case, consider an engineering student with two years of mathematics who has decided to switch. His quantitative background is apt to be so impressive to his adviser that he will be excused from any possible requirement for a special mathematics course. What might be useful for all these students is some special junior-level course; a freshman course for the social sciences is not apt to be an ideal solution.

But that is getting well ahead of my task. What the data show is that so far as freshman courses are concerned, what counts most is not the final degree, but the initial degree plan.

Let us look at enrollment figures for the Fall of 1980, courtesy of CBMS, in certain selected courses at the level of calculus or below.

Table 3. Enrollments in selected elementary courses, Fall 1980

Course	Enrollment
Calculus for EMP	405,000
Calculus for social science, biological sciences, management	104,000
"Business math"	48,000
"Finite math"	95,000
Probability without calculus	13,000
Total enrollment for these courses	665,000

There were a total of 1,368,000 enrollments (including those in Table 3) in mathematics courses of the level of calculus or below. I have listed every one that is not remedial, or in the traditional pre-calculus sequence of college algebra, trigonometry, analytic geometry, or designed for elementary teachers or for a "liberal arts" course. That is, these are all the mathematics courses at this level which could form part of a degree program outside of EMP, or which might conceivably be suitable for a computer scientist. If we compare these totals with 1980 freshman plans, we find we have reached not very many of these special groups.

Table 4. Some prospective freshman majors, Fall 1980

Major		Number Planning
Biology		304,000
Management, business		409,000
Social science		115,000
	Subtotal	828,000
EMP		330,000
	Total	1,158,000

Not everyone in this conference has taught calculus lately, and some unfortunates may never have done so. We have now nearly gotten back to the level of Granville, Smith and Longley, and are still sinking. If I speak of a course as having the difficulty of calculus, I am not talking about a very demanding, highly rigorous course.

I will say something here about enrollment in those mathematics courses past calculus suitable for majors outside EMP, postponing discussion of computer science. There is not much available. That is not necessarily bad. There were fewer than 1,000 students taking Biomathematics, for instance. Considering that only 2% of university mathematics departments, 8% of public college departments, and 1% of private college departments offered such a course, the small enrollment is inevitable. But does that <u>prove</u> that mathematics departments are neglecting the needs of the biology major? They may be, but these figures do not demonstrate that. The biology major planning graduate work in, say, population dynamics, needs a great deal of the same sort of mathematics as the engineer does, taught perhaps with a different emphasis ("compartmental analysis" in differential equations, for example). But one cannot tell from such figures if this changed emphasis actually happens.

Table 5. Enrollments in selected post-calculus courses, Fall 1980

Course	Enrollment
Linear algebra/matrix theory	37,000
Applied math/ math modelling	2,000
Biomathematics	1,000
Operations research	2,000
Probability (calculus)	13,000
	55,000

The linear algebra course figures are for all linear algebra students. There was a total of 184,000 students enrolled in post-calculus mathematics courses.

In the Eighteenth Century, there were performed many strange metamorphoses of Shakespeare's plays. In none of these, I think, did Hamlet first appear in the third act. Some readers of this document must feel somewhat as though they were in the audience at such a production, however, and think it is very late for computer science to appear.

We examine first the change in enrollments in mathematics, computer science and statistics, summarized from the CBMS Report [1]. We see in Table 6 that mathematics and computer science enrollments have both grown from 1975-80 to 1980-81, mathematics by 274,000 and computer science by 189,000. But the percentage increases are very different; for mathematics it was 22%, for computer science 185%. In both fields the great bulk of the teaching is elementary, 89% in mathematics and 79% in computer science. In mathematics almost all the growth occurred at the level of calculus or below; enrollments in post-calculus courses went up only 5%. Mathematics has more than 5 times the enrollment of computer science; but the two fields are in quite different stages of development.

In any consideration of new courses or of changing topics from one type of department to another, the question of faculty size becomes crucial. From the CBMS Surveys, we have quite good figures on faculty in mathematics proper, but I believe they are less reliable, though not

Table 6. Course enrollment (in 1,000s) in undergraduate mathematical science courses, 1970-1980, in universities and four-year colleges

Subject Group	1970	1975	1980
Mathematics:			
Remedial math	101	141	242
College algebra/trigonometry	301	259	345
Analytic geometry/calculus	345	397	518
Other pre-calculus*	228	292	250
Subtotal pre-calculus	975	1,089	1,355
"Useful" advanced**	129	97	131
Other advanced	111	66	40
Subtotal advanced	240	163	171
Subtotal mathematics	1,215	1,252	1,526
Computer science:			
Elementary	n/a	85	230
Intermediate	n/a	12	43
Advanced	n/a	5	18
Subtotal computer science	79	102	291
Statistics and probability:			
Elementary	n/a	99	104
Intermediate	n/a	22	29
Advanced	n/a	20	16
Subtotal statistics and probability	90	141	149
Total, all mathematical sciences	1,384	1,495	1,966
Total freshman enrollment	1,617	1,761	1,712

*Other pre-calculus courses are business math, liberal arts math, math for elementary teachers, finite math.

**"Useful" advanced courses are differential equations, linear and matrix algebra, advanced calculus, applied math, numerical analysis, all courses frequently taken by non-math majors.

bad, for statistics and computer science. I give in Table 7 the figures for universities, public colleges and private colleges.

Table 7. Mathematical science faculty and teaching assistants in universities and four-year colleges

Type of Department		1970	1975	1980	1980/1975
Universities:					
Math.	full-time	6,235	5,405	5,605	3.7%
	part-time	615	699	1,038	48.5
	t.a.'s	5,999	5,087	5,491	7.9
Stat.	full-time	700	732	585	-30.1
	part-time	93	68	132	94.1
	t.a.'s	747	690	546	-30.9
C.S.	full-time	688	987	1,236	25.2
	part-time	300	133	365	74.4
	t.a.'s	309	835	1,813	117.1
Public colleges (4-yr.)					
Math.	full-time	6,068	6,160	6,264	1.7
	part-time	876	1,339	2,319	73.2
	t.a.'s	1,804	1,805	1,535	-15.1
C.S.	full-time	n/a	n/a	436	-
	part-time	n/a	n/a	361	-
	t.a.'s	n/a	n/a	90	-
Private colleges (4-yr.)					
	full-time	3,352	3,579	4,153	16.0
	part-time	945	1,359	2,099	54.5
	t.a.'s	146	559	1,154	106.4

Comment: In 1980, over 25% of all t.a.'s are _not_ mathematical science graduate students, up from 6% in 1975. Presumably, most of these are in math departments.

A word or so of interpretation, first: (1) The CBMS Survey Committee definition of "college" depends on the number of separate schools in the institution. The University of Illinois at Chicago Circle, with many strong graduate programs, is a four-year college; so is Slippery

Rock State College. Marquette University and Mary Crest College are both private colleges. (2) For both statistics and computer science, figures are given only for departments with names clearly indicating their task. Three statisticians in a psychology department would not be counted, though in another university they might well be in a statistics department. Very few private colleges have separate departments either of computer science or of statistics. Such faculty members are not here counted in their fields.

In mathematics we feel overworked because the increase in our load has not been accompanied by a corresponding increase in faculty. In universities (as opposed to colleges) the full-time mathematics faculty went up only 3.7% from 1975 to 1980 while enrollment went up 22%; the full-time computer science faculty went up 25.2% to cover an 185% increase in enrollment.

A 250% increase in advanced course enrollment in mathematics in the next five years is improbable, but could be rather easily handled. A 250% increase in computer science intermediate and advanced undergraduate enrollment may not be handleable at all, though the demand is very likely: We know that in 1979-80, 83,900 freshmen planned to be computer science majors. Table 8 I find very instructive; it comes from ACM data [3].

Table 8. Ph.D. production and Faculty per U.S. Ph.D. awarding computer science department

	1975	1979
No. Ph.D. awarding departments	60	77
Ph.D. grads per department	4.3	3.2
Total Ph.D. degrees	258	249
Faculty per department	14.6	12.4
Total faculty	876	955

The total production of Ph.D.s actually shrank. In those five years, there were approximately 1,250 new computer science Ph.D.s; the graduate departments got 70 (at most). If the department faculty had increased as the enrollment, the total faculty would be 2,500; the Ph.D. awarding departments could have hired <u>all</u> the new Ph.D.s.

There is a discrepancy between Table 7 and Table 8 that worries me. From Table 7, the full-time university computer science faculty rose 25% in 1975-1980. From Table 8, the "total faculty" of Ph.D. awarding departments grew 9% from 1975 to 1979. I can think of a number of possible explanations. However, even the larger increase is so small as compared with the growth in enrollment that conclusions can hardly affect the purposes of this conference.

Except that it is large, it is hard to make predictions of the need for bachelor's level computer scientists, and I distrust any I have seen. Other ACM data [3] estimate that in 1981 there were 1,771,000 computers in use, and that by 1984 there will be 4,363,000. This presumably means an inexhaustible demand for at least short courses in programming. But it also puts tremendous pressure on computer manufacturers for user-friendly hardware and software, and on CAI experts for computer-taught programming, on the job, all forces against the need for majors.

What I think we will see is a continued expansion of elementary teaching in computer science. We will not need as many bachelor degree programmers in the next decade as are predicted. But we will need many--I can't quantify further--sophisticated computer scientists with a background that I regard as mathematical, whether my fellow-mathematicians or the computer scientists will agree. (There will undoubtedly be needs for many other types of computer scientists.)

Table 9 gives some existing courses and their 1979 enrollments that illustrate the sort of subjects I have in mind. The mathematics course enrollments total 51,000. It seems to me improbable that as many as 10,000 of these are computer science majors. All the mathematics courses listed normally have a calculus prerequisite.

It seems to me that the courses in Table 9, or ones very similar, will form the basis for most theoretical computer science. One wonders how many computer science departments wish to add to their offerings of these courses, given their desperate manpower situation. And for all the courses which now have as prerequisite courses in the standard EMP sequence, one wonders how desirable or necessary these prerequisites are.

Table 9. Courses with mathematical content for computer science and 1979 Fall enrollment, in 1,000s

Course	Enrollment
Mathematics:	
Linear algebra	37
Modern algebra	10
Combinatorics	1
Set theory	1
Mathematical logic	2
Computer science:	
Discrete structures	9
Data structures and algorithmic analysis	12
Algorithms	2
Automata, computability, formal languages	2
	76

From a purely pedagogical standpoint, given my own intellectual background, I would find it much easier to teach set theory, for example, to students all of whom have had the elementary real variables found in calculus and advanced calculus, and there must be hundreds of mathematics faculty who would agree. We have not had to face the problems of teaching such courses to students with radically different backgrounds, even though just as "good" mathematically.

The problems in the last several paragraphs have, of course, been much on Tony Ralston's mind; he is the first person to have pointed out their implications for elementary mathematics. All I am doing is saying that the numbers support his position.

One final table, and I am done. As a result of the computer, the power of mathematical methods has increased greatly. It would seem obvious that students should learn to use this new tool in mathematics as early as possible. Yet that is not happening. I call your attention to Table 10 derived from the CBMS Survey [1], which shows how few students are using the computer in their mathematical courses.

Table 10. Percentage of students in universities and four-year colleges using the computer, by level of course

Course	Percentage of Students
Mathematics:	
Below calculus	3
Liberal arts math	12
Calculus	3
Upper level	6
Linear algebra	12
Applied math	16
Statistics	21
Computing and related math	83

Bibliography

[1] Conference Board of the Mathematical Sciences, Report of the Survey Committee, Vol. VI, Undergraduate Mathematical Sciences in Universities, Four-Year Colleges, and Two-Year Colleges, 1980-81. CBMS, Washington, 1981.

[2] Modified from Projections of Education Statistics to 1987-88, Department of Education, and cited in [1].

[3] Peter J. Denning, Eating our seed corn, Commun. ACM, vol. 24 (1981), pp. 341-343.

DISCUSSION

Young: The main point I would like to emphasize is that vast numbers of people are studying mathematics who are not going into traditional fields that use mathematics. The process is more or less out of control. For example, students majoring in the social sciences often make their decisions on a major late; thus, their mathematics background at the end of their sophomore year is what they are stuck with. As another example, you don't get a degree in range management at the University of Wyoming without a year of mathematics.

Tucker (Albert): I am quite optimistic about the possibility of change. Calculus used to be a sophomore course and could be again given only that the first year contained some elementary aspects of calculus needed by other disciplines.

Lochhead: Although many physicists might not agree, I think that a lot of what physics needs out of mathematics can be accomplished using computers (such as at Dartmouth) to get students to understand and work with various functions. Also I believe that introductory physics makes a terrible mistake in covering far more content than students can possibly absorb.

Tucker (Alan): In support of this the CUPM Mathematical Sciences report recognized that applications do not necessarily come after theory so that there is nothing intrinsically wrong with a physics course laying the foundation for mathematics before the mathematics is taught. On the Mathematical Sciences panel many people felt that calculus should be a sophomore course but we didn't have the guts to say this.

Barrett: We should remember that mathematicians talk a lot about the beautiful unity of the calculus course but few students see this until, at least, they have taken some more mathematics.

Kreider: There is a tendency to think that calculus is a beautiful block of material and that, if you chip anything out of it, then it is no longer calculus. But, as Al Tucker noted, some of the concepts and topics relevant to other disciplines can be woven into a first year curriculum and the "block of beautiful material" could well be studied later on.

Barrett: I'm bothered about being in a school that does not have an engineering college but where the ideas of engineering and physics still dominate and where some mathematicians believe that only the "real" calculus is a good course to teach.

Bushaw: I want to come back to the point Gail Young raised about the hordes of unidentified students who take our beginning courses for no clearly understood reasons. Our department, for example, tried to find out why students in Forestry are required to take our pre-calculus course but couldn't. I suspect, as Dick Anderson has suggested, that it is mainly a filtering device.

Roberts: About ten years ago at Rutgers we tried to design a mathematics course for biologists which would include some probability and statistics and discrete mathematics as well as calculus. But the biologists resisted because they didn't want to be made to feel like second class citizens. This is an attitude which we're going to have trouble with.

Symbolic manipulation and algorithms in the curriculum of the first two years

Herbert S. Wilf
University of Pennsylvania
Philadelphia, PA 19104

1. Introduction

Because of computers, something old can go and something new is coming. That much is certain. The disagreements begin when we start making lists, but in the interest of provoking debate, I'll take the plunge. This paper is divided into two main portions. First we'll discuss the impact on the curriculum of computers that do symbolic manipulation. These were originally only large mainframe computers. Now, as described for instance, in [1], small personal computers can do it too. The next stage of the revolution will occur when such capability is available on small portable "pocket computers", and so we will talk about how the classroom environment might change at that time.

The other portion of this paper concerns the teaching of algorithms in the early years of college. The rapid surge of interest in computers and computing has, in my opinion, brought algorithms into the select arena of those subjects that must be within the ken of the Educated Person. We will discuss here exactly which parts of the theory and practice of algorithms fit into the early curriculum and how.

2. The impact of symbolic manipulation hardware.

What will have to change, in the current standard mathematics program, falls roughly speaking into two categories. First, certain portions of the curriculum presently teach concepts or methods that computers will soon be able to handle even better than people can. Of course, just because a computer can do it doesn't imply that people shouldn't learn it. But some things are of that type: not only can computers do them better, but they have little or no redeeming intellectual value, so why not relax and let the machines carry the ball?

Secondly, the pressure of the ideas that must enter the curriculum is, in certain cases, so intense that some topics will have to be deleted just to make time available for the persistent newcomers. In other words, even though the dropped portions may retain a good deal of value, they won't be able to compete with the thrust of some of the new ideas that are screaming for attention.

Let's begin with some areas of least disagreement, hopefully. What computers can do and should be allowed to do and people's heads shouldn't be bothered with routinely are those tasks that are strictly mechanical, fairly tedious, without redeeming social merit, etc.

Now take a "cookbook" course in differential equations, for instance. Typically we teach a variety of special forms of differential equations that can be solved analytically, we show the special devices that allow us to get the closed-form solutions, we illustrate the methods both with word problems and with sample differential equations, and so forth.

Suppose, to continue a theme raised in [1], that I can buy at my local 5&10 cent store a little calculator that, for $39.95 say, will handle differential equations. First we push the "DIFF" button, to put it into a frame of mind for solving differential equations. Then in response to the prompt "ENTER EQUATION, PRESS END WHEN FINISHED" that appears in the LCD window, we key in $y'' - 4y' + 4y = 0$ and hit END. Next we get a prompt "INITIAL DATA? (Y/N)" and we respond "N" because we happen to want the general solution rather than a particular one. The thing thinks for a few seconds, and then halts with "(A + B*x)*exp(2*x)" in the display.

For a more striking example, consider an inhomogeneous equation, say the equation

$$y'' - 4y' + 4y = x^3 + 3x^2 + x^2e^x$$

We know that each of the terms on the right side has a finite family of derivatives, so "all" we need to do is assume a solution in the form of a linear combination of all of the members of that family, and then "just" solve for the constants in the linear combination. This routine process is quite messy, the chances of error are high, and our $39.95 computer (well, maybe $139.95!) should be able to do the job symbolically, quickly, painlessly and correctly (thus proving its superiority, as a life form, over the author of this article, who can do the job symbolically, but none of the other three adverbs apply).

With these thoughts in mind, I ran through the table of contents of one excellent book that serves as the text in second year calculus in many colleges and universities, and I made a list of some of the topics that are often taught, and that could in principle be done on a little symbolic calculator of the future. Here is part of that list.

Under infinite series, we will be able to enter a function symbolically and get its Taylor series coefficients about a given point as rational numbers; enter f symbolically, and then, each time DERIV button is pressed, display the derivative of the function previously displayed;

Under complex variables, the calculator will do complex arithmetic, find argument and absolute value, find powers and all roots, calculate residues of a suitably factored rational function at a point, evaluate the integral of a rational function over the real axis if its denominator is factored, find the factors in question if they involve rational numbers, etc.

With matrices, expand determinants as rational numbers; do matrix algebra, including inverses, rationally; compute reduced echelon form of a matrix in rational arithmetic, and so find the rank and nullity of linear systems; find the general solution of a system of simultaneous linear equations, in the form of a particular solution plus a free linear combination of a set of basis vectors for the kernel (no real numbers used, no roundoff errors, arbitrary constants appearing as literals in the screen display, just as we would normally write it out);

3. Discussion

The capabilities described above are well within the powers of present day microcomputers with their disk storage and 256K random access memories. The next stage in the evolutionary process, then, will be the increasing availability of symbolic manipulation packages of greater sophistication on widely available microcomputers.

Also, I must remark that I don't see anything threatening in any of the above developments. What computers are doing is that they are gradually teaching us what people are. It can be unsettling to realize that what we previously thought was a very human ability, like spot welding, for instance, or like calculating the residues of a rational function, for another instance, can actually be better done by "machines". Following that first unsettled reaction we would probably agree that after all, it isn't terribly exciting to calculate residues, and it's just as well that calculators can do it for us.

Imagine the enhancement of the learning process that would come from having the students use their calculators in class; after all, aren't we always looking for good classroom demonstration materials?...

> *"Now today we're going to study Taylor's series. If you'll all take out your calculators, let's find the expansion of the function f(x) = ln(1+x) about the origin. First key in LN(1+X) to the display. Next let's all watch what happens as we repeatedly hit the key DIFF followed by the key EVAL. After DIFF the display is replaced by the derivative of what used to be there. Next store the new function by hitting STORE F1. Then evaluate it at the origin by keying EVAL X 0. Now you're looking at f'(0). Print it. Now recall the first derivative from memory with RECALL F1..., and around we go again. Hit DIFF, then..."*

The lecture room scenario above represents my own personal wish for how the impact of symbolic manipulation hardware should get absorbed into the educational system. In other words, I very much hope that the calculators become thoroughly integrated into the course material, so the student will come away with a knowledge both of the conceptual substance of the course and of how to use his/her calculator to get the desired answers. The two modes of instruction will surely complement each other: use of the calculator will make the classroom material seem much more real, and the class work will reveal to the students the immense power of their gadgets.

4. Description of the gadgets.

I think it may be helpful to have a clearer picture of what capabilities a little calculator might have, in order to focus the discussion a little better.

So please join me in imagining something flat, about 3" by 8", with a long LCD window for display, and lots of buttons to push, including a complete alphabetic character set, a complete numerical set, and many special functions. Among the special functions will certainly be the familiar ones of arithmetic, as well as the standard elementary trigonometric and power functions.

Beyond that we will be able to clear the display, then enter, in the usual alphabetic notation, some function, and by pressing the STORE key followed by the name of one of the function registers, the function will be saved. The DERIV key, when pressed, will cause "WITH RESPECT TO?" to appear in the display, and if we then key in the variable name, such as X, T, etc., the requested (partial) derivative of the function formerly in the display will appear in the display, as much of it as fits, that is. A SCROLL key will cause the display to scan the entire expression.

The EVAL key, when pressed, will cause "REPLACE WHICH VARIABLE?" to appear, and when we answer "X", for instance, "BY WHAT?" will appear next. Then if we enter "2", the value of the original expression in the display, with every appearance of x replaced by 2, and simplified as much as possible, will appear (imagine doing a problem in L'Hospital's rule with this computer, for example).

The INTEGRATE key also brings forth "WITH RESPECT TO?" when it is pressed, and when we answer, the calculator will do its very best symbolically to integrate the expression that was in the display. No doubt early models will have relatively few strings in their bows, but it won't take too long before very comprehensive integration ability will be store-able in very small amounts of hardware.

The MATRIX key will tell the computer to put on its linear algebra hat. It will respond with "NAME OF MATRIX?", then after we enter "A", or some such name, it will ask "NUMBER OF ROWS?", and then for the number of columns, and finally, it will let us enter the matrix elements one at a time, in some standard order, separating elements with the ENTER key. These entries will be stored in the matrix registers as they are entered, and will be thought of as the matrix A (in later models, the entries might be allowed to be themselves symbolic, though $39.95 might then be left far behind!).

When finished, we might enter another matrix B, if we wished, preparatory to asking for some matrix arithmetic, such as multiplication, or addition. Otherwise, maybe we will want the solution of the linear system whose augmented matrix we have just entered. One push of the RREF key causes our little wonder to think for a while, and then halt with the reduced row echelon form of the input matrix in its head. From this we will be able to answer virtually any question about the given linear system, with no roundoff errors because all calculations will have been done in rational arithmetic. This one RREF capability will allow us to find rank, inverse, solution vector when unique, basis for solution space and rank when not unique, and determinant when input is square.

Finally there's the differential equation capacity that I've already referred to: finding symbolic general solutions or particular solutions with given initial data, of wide classes of differential equations, including first order linear, first order separable, equations with constant coefficients if the roots of the associated polynomials are rational, and the same with right hand sides like polynomials, exponentials and other familiar functions.

It has been suggested to me that perhaps the effect that I am describing would already be available with good audio-visual materials or with a microcomputer arranged so that the whole class could watch the display. I must disagree. Already we have seen some enhancement of the learning process when we replaced remote batch-fed computers with small independent micros, one per student. The student can identify with the microcomputer as something that he/she might own, and so time spent on it is easy to justify as an investment in the future.

Even more so, I think, will portable calculators enhance the impact. Being portable lends to such calculators the effect of being a part of the personal capability of the user. That is, skills we acquire that can be used on a little calculator seem more like our own skills, and have an immediacy that might otherwise be lacking.

Though it may be unchivalrous of me to mention it, there is also the practical point that many instructors regard A-V setups as time-consuming hassles that fail to be cost-effective, and so they don't use them.

Finally, how much, by volume, of the first two years of the mathematics curriculum could be replaced, if such symbolic aids were widely available? The first year seems surprisingly tight. Surprisingly, because when I scanned it to see what could go I was expecting to find large quantities of unnecessary busywork. But that isn't the reality. Students are taught integration by parts, for instance, and if just one homework assignment is given on that subject, well, that isn't excessive, calculators or not.

In the first year we probably could cut the total time spent on methods of integration by half, so that each method gets just a quick once-over-lightly, two problems to do without calculators, and a few to do with calculators. With some other small economies of this kind, having to do with drilling skills, I would estimate that the first year course could shrink by about 15%, losing only bath water, and keeping the baby. The second year could be cut perhaps by 1/3, reflecting the usefulness of the calculators in doing the manipulation parts of the work in differential equations and linear algebra.

Naturally I am full of suggestions about how all of that time might be profitably used, as described in the next section.

5. Algorithms and the early curriculum.

The discussion of symbolic manipulation hardware has a bit of pie-in-the-sky about it because that hardware just isn't for sale yet, in the form of pocket calculators.

In contrast, the subject of algorithms is here to stay, and has already established itself as one of the most important new subject areas of our time. What are the core ideas, the ones that deserve a place in every liberal arts or technical environment?

First, in my opinion, is the concept of a formal algorithm, as opposed to a "first do this for a while and then this", or informal approach. The precise organization of a binary search in an ordered list of integers to see if a certain integer is or is not present is a skill, or an appreciation of one, that our proverbial Educated Person can't be without. Informal algorithms are, of course, better than nothing. But somehow, to be in tune with the burgeoning Computer Age, it seems to me that instruction in formal algorithms is necessary. Furthermore it should happen early if the student is to make an intelligent and informed choice about a major subject that may involve computers in some way.

Second, beyond the concept and execution of formal algorithms, there lies the concept of a recursive algorithm. Now we're in deeper water. All computer scientists will recognize this subject as essential, but is it really *de rigeur* in a liberal arts curriculum?

I suggest that it is. The idea of mathematical induction is already one of the most important in mathematics, even though its mastery is not required of all students. When we consider the confluence of the fundamental role of the idea in mathematics with its equally fundamental position in computer science, it seems to me that the case is overwhelming.

First we might have the usual bowling-pin introduction to mathematical induction, with some examples of its power in proving theorems. Then we continue beyond that to the use of induction in formulating definitions:

> *Let's define the gcd of two nonnegative integers a and b. First, if b=0 then the gcd is a. Else, if b is not zero, the gcd is the gcd of b and a mod b.*

Have we really defined something? Or have we fallen into the cheap-pocket-dictionary trap: a being is a thing and a thing is a being? Once the examples of the factorial of n and of the gcd have been mastered we can move on to a few of the elegant and deeper uses of recursive algorithms, such as Quicksort and Chromatic Polynomial, for instance.

In Quicksort, we are given a list of n numbers, and it is desired to sort them, i.e., to arrange them in ascending order of size. To do this by Quicksort, we would do

(a) if the given list has no more than one element then output the input list, untouched

(b) else, switch the numbers around a bit so as to create one element of the list, the "leader", that is bigger than everything before it and smaller than everything after it in the list.

(c) Quicksort the elements before the leader, and Quicksort the elements after the leader. Done.

It is certainly a Major Idea of Western Man, etc. etc., that the algorithm is now well defined (or at least it would be if I had described how to "switch the numbers around a bit"!). It is, furthermore, an extremely important insight for students to realize that in many high level computer languages the compilers can deal with a recursively described program directly, without any need to make it nonrecursive first. I believe that all of these concepts deserve a place in the curriculum of the first two years.

Now I've mentioned the idea of a formal algorithm and the idea of a recursive algorithm. There is one other concept that deserves the same special attention, and that is the notion of computational effort. To teach this we would discuss the various measures of effort, such as space, time, programming complications, etc., then analyze the running time of some elementary algorithms, then discuss the polynomial-exponential-beyond hierarchy of time complexity (using the areas of the subject where the algorithmic improvements are of practical as well as theoretical importance), and finishing with a few examples of "hard" computational problems, i.e., ones where exponential computing times are the best known.

A classroom discussion of algorithms could not be complete without hands-on implementation of the algorithms on computers by the students. Hence we presuppose the ready availability of an extremely simple introductory language, like Basic, and a still-simple but recursive language, like Pascal or PL/C, to provide the necessary experience.

The three concepts, of formal algorithm, recursive algorithm, and computational effort, deserve, I think, an assured place in the core of the undergraduate educational process in a liberal arts institution.

6. The role of algorithms in the curriculum.

If we can agree, at least for the sake of discussion, that algorithms do belong in the introductory curriculum, there remains the little question of how this is to be done.

After the development of the calculator hardware that I spoke of earlier, in the mass-market price range, there will be room in the mathematics courses themselves for the introduction of algorithmic ideas and for hands-on practice in using those ideas. In the transitional period, before symbolic manipulation hardware becomes widely available, a separate course will be needed.

At the University of Pennsylvania, for example, we are now trying out a computer laboratory for freshmen or sophomores. This course will provide 1/2 credit and will teach the writing of programs in Basic and in Pascal, implemented on our microcomputers in the Mathematics Department.

This may or may not be ideal. But at least it addresses the right question, of how to make available, very early in a liberal arts curriculum, in a University that is very proud of its liberal arts tradition, exposure to the concepts and practice of computing.

To help fix ideas, I will now suggest a syllabus for a possible one semester course, to be given in the second semester of the sophomore year. This course would serve during the transitional period, i.e., until super calculators become widely available (numbers in parentheses are the numbers of class meetings that might be devoted to the suggested material):

> *Introduction to the local computer system, to the Basic language and to the Pascal language (6);*
>
> *Numerical algorithms: Bisection method, Newton's method for square roots, Newton's method for polynomials in general (5);*
>
> *Formal algorithms: how to search, by bisection, an ordered list; informal statement, formal statement, and preparation of Basic or Pascal programs; analysis of the speed of the algorithm (4);*
>
> *The Euclidean algorithm for the gcd; write program in Basic; study worst case behavior; generalize to gcd of many integers (3);*
>
> *Restatement of Euclidean algorithm in re-*

cursive form; write Pascal program; Discussion of Quicksort: statement as formal recursive algorithm, write Pascal program (7);

Introduction to the complexity of algorithms: analysis of speed of matrix multiplication (traditional method); description of Strassen's method of multiplying 2x2 matrices, and of the consequent general matrix multiplication program; formulation of the latter as a recursive formal algorithm; analysis of speed; programming in Pascal optional as a special project (7); discussion of how large matrices have to be before a real speedup occurs.

Some exponential-time algorithms: write an exhaustive search algorithm for the travelling salesman problem; estimate numbers of cities for which the method is practicable; compare results with the greedy algorithm (3);

Discuss bin-packing; write first fit decreasing program for bin-packing problem, using Quicksort prepared earlier in the course (2);

A seemingly exponential time problem that isn't: minimum spanning tree; write program and analyze complexity (3);

In remaining time, if any, assign special projects to students, each involving a reading assignment combined with writing, testing and documenting a program.

After the right hardware becomes available I would expect a re-thinking of the standard freshman-sophomore mathematics courses to take place. My personal vote would be in favor of replacing portions of the present curriculum by an introduction to algorithms and computers. The portions that would be replaced would be first the average 20% or so that I spoke of earlier as being do-able better by smart calculators, and second perhaps another 10% that, while still needed, is just plain not as urgently needed as the algorithmic and computing material.

Not only would the material on algorithms be presented in a separate part of such a course, the mode of thought that that material engenders would enhance even the standard portions of the subject.

The analogy of functions and subroutines mutually reinforces the learning of both of those ideas. The independence of the function itself from the letter that symbolizes its argument will be more meaningful to anyone who has looked into global and local variables in a programming language, and conversely, the former concept strengthens the latter. Another example would be the ideas of a summation sign and a for...next loop, and so it goes.

The prospect of courses emerging in which busywork is much reduced, and in which modern algorithmic ideas are present, both for their own sake and for enhancement of the mathematical concepts, is to me extremely attractive. I look forward to the debates that no doubt will accompany the developmental processes, and to the fruits of those debates which, it seems to me, can only be a significantly improved educational experience for the students and for ourselves.

Reference

1. Herbert S. Wilf, *The disk with the college education*, American Mathematical Monthly, Vol. 89, No. 1, January 1982, pp. 4-8

DISCUSSION

Wilf: On the symbolic manipulation part of my paper the point I want to emphasize is that the importance of the availability of symbolic systems on microcomputers is not so much in what can be done today but in what such systems portend for tomorrow.

Bushaw: The official position of groups like the National Council of Teachers of Mathematics (NCTM) is that, in the elementary grades, calculators should be used and the curriculum should be modified to acknowledge their existence. I take the same view about symbolic manipulation in higher levels of the curriculum.

Pollak: There are 10^6 elementary teachers in this country all of whom have to teach some mathematics. What really ought to be done is to get every one of these 10^6 people to an institute for a week which would take the fear of calculators away and get rid of the feeling of immorality that is associated with calculators.

Anderson: I very strongly favor greatly reduced student practice in the formal several-digit algorithms of arithmetic. They're inevitably going to be little used in the computer age. But I don't yet understand how deemphasizing algebraic manipulations by students will affect learning and understanding and performance in mathematics.

Maurer: I think we still want to encourage people to have some memory so that they can remember a few simple things. More important perhaps is to have or develop mathematical capabilities for estimation when the machine is not available or useful.

Young: In a letter written in the 12th century by a mathematician to a German merchant about his son's mathematical education, it is said that, if you want to learn addition and subtraction, you can do that in our German universities but, for multiplication and division, you must go to Italy. The point is that we're faced with the same sort of obsolescence of the training of university mathematics faculties. As an example, it serves no purpose for us to sit here and say that students should learn how to do partial fractions by hand. They're not going to do it that way.

Barrett: I feel very strongly that mathematics is an experimental science. Thus, I favor letting students experiment with a calculator or computer before they are taught the formal mathematics.

Weissglass: We need to make a distinction between technical proficiency and understanding. When we start thinking about a new curriculum, we must also think about the mechanism of change and how the method of teaching will be changed.

Wilf: I'm hoping that at this conference we can express some of the liberated feeling achieved by symbolic systems and try to find out what we would like to replace some of the techniques by.

On the second part of my paper, the role of algorithms in the curriculum, let me emphasize:

- the importance of understanding the limitations of computers and algorithms, that is computational complexity at a low level
- the need to teach students to think recursively and
- the desirability of expressing algorithms formally enough so that they can be fully understood and analyzed.

Tucker (Albert): We must be careful about computational complexity because it is still at a very crude stage. But more generally I would point out that algorithms should play a more important role in continuous as well as discrete mathematics. As an example, there is practically no textbook in linear algebra that gives an algorithm for determining linear independence.

Anderson: I'm concerned about the use of the word algorithm. One of the defects of school mathematics is that people only memorize algorithms. We need to teach not only algorithms but comprehension of them.

Norman: Computational complexity may not be quite the right approach but I am in favor of teaching what a computer or algorithm can and can't do and in comparing algorithms for the same task.

PROBLEM SOLVING AND MODELING IN THE FIRST TWO YEARS

by
William F. Lucas
Cornell University
Ithaca, NY 14853

1. INTRODUCTION.

Mathematical education is currently confronted with a variety of substantial problems. This paper discusses three of these problems which are related to the role of finite mathematics and its appropriate place in the undergraduate curriculum. First, there is the pressing problem of how to incorporate important new mathematical subjects and discoveries into an already crowded curriculum, including discrete topics at the elementary level. Second, there is the recently emerging problem that the mathematical community, except for computer science, is very rapidly losing its base of talented young people. Third, there is the question of how to design courses which give due attention to the revolution and needs in discrete mathematics and are worthy enough to compete with traditional calculus courses. It is important in addressing this first problem of the "curriculum crunch" to not overlook the second problem of "diminishing base" which could be devastating to the health of the general mathematical sciences if allowed to continue for long. It is also not clear to many that we have yet come up with the appropriate finite mathematics courses to take a major slot in the freshman-sophomore years of college.

Two of the major reasons why mathematics, minus computer science, has rapidly been losing student interest is that we have stripped the "fun" out of doing mathematics and we have furthermore neglected to convince students in a meaningful way that mathematics is a contemporary, dynamic and currently relevant subject. If the mathematical community makes significant curriculum changes without attention to such "fun" and "currency" they will risk aggravating the problem of the diminishing base. E.g., in attempting to cover more finite mathematics earlier in the college curriculum, it would be a shame to not introduce exciting problem solving and realistic modeling from relevant modern applied areas. A wealth of new material is now available to provide highly motivating and interesting topics while simultaneously meeting other various objectives and goals for such new courses at the freshman (and lower) levels. Whether there should be a one or two year discrete mathematics sequence seems open for further consideration.

2. GROWTH.

The problem of fitting more discrete mathematics into the first two years of college mathematics is a special case of a more general pressing problem. There has been an astonishing growth in the amount of new mathematics in recent decades, and a significant part of this enormous expansion is relevant to mathematical education at the college level. This includes several developments of truly revolutionary impact such as digital computers and related theory, the advent of finite mathematics on a large scale, the spread of mathematics into so many new areas of theory and application, the resulting new intuitional bases for mathematical discovery, as well as several other factors. These newer fields are not passing fancies nor disjoint curiosities, but are here to stay. They will find a place in the undergraduate curriculum somewhere, either within mathematics or else in some other departments. Although this expansion is a very healthy thing for the general state of mathematics itself, it has created the difficult problem of deciding what to include in the mathematics curriculum, and when. It is becoming increasingly impossible to squeeze all of the desired subjects and topics into a four year program, and some sort of compromise or earlier specialization will be essential. The extension to more professionally oriented master's degree programs may help to take off some of the pressures caused by the "curriculum crunch".

There are a few changes of a more numerical or administrative nature which could bring some slight relief in the squeeze on the college mathematics curriculum. These include steps such as the following. The first two or three semesters of calculus, or any precalculus courses, could not be counted for credit towards the major. This would be similar to the physics majors who take calculus without credit in their major. This could also be done for one or two basic computer science courses as is common at many schools. Perhaps history or broad survey type courses within mathematics could be counted toward requirements other than in the major. Above all, it is becoming abundantly clear that serious majors in the mathematical sciences should be taking two courses in each semester in the general mathematical sciences area during the first two years of college. It makes little sense any longer to delay courses on computer science or finite mathematics until upper division studies, no more perhaps than to delay calculus until then. If mathematics were restructured into a division with a few departments within it, then one could be taking math courses in these different departments. However, this latter would apply only to large universities. Nevertheless, such counting tricks to add a couple of extra mathematics courses is hardly enough to solve the major problem.

The question is not so much whether important new finite mathematics will be taught at lower levels. It surely will. Either within mathematics or elsewhere, either to lower division or precollege students, and either poorly or well. If there are no simple "administrative" ways as discussed above to squeeze it into the curriculum, then some more sacred notions about the mathematics major will have to be altered. This will include notions about what the undergraduate core is (if not empty) as well as reordering of the usual mathematics courses. We cannot maintain a large rigid core mainly because it currently exists, or because some have admonished us to the effect that no part of it can be "safely neglected" by even those exclusively interested in the applicability of mathematics. Another possibility will be that students must select and specialize earlier, despite the distastefulness of this to many. (Let us not get carried away here with excessive concern with "premature decisions" or the "transfer students". Students will already have made more major decisions.) Still another possibility is the teaching of broad introductory college courses which tend to be more integrated and survey in nature. Perhaps we can learn from those in the physical and biological sciences who confronted the squeeze some time ago and do not seem to be overwhelmed by it. They do not seem to be having major battles over slots or time in their curriculum.

3. WHERE HAVE THE STUDENTS GONE?

Any consideration of changes in mathematics curriculum at the college level must seriously address another important problem which has arisen in recent years. There is a rapidly diminishing base of well qualified students at the upper class college and graduate school levels. Except for computer science which is still expanding, mathematics students are voting with their feet and deserting fundamental upper division mathematics courses and majors. This is clear from enrollment figures such as given in CBMS surveys. A glance at the December Notices of the AMS indicates how few bachelor's degrees in mathematics are being awarded by most listed institutions. (It is true that the Notices do not cover all schools, and that it also shows a few schools with very popular mathematics programs which should be carefully studied to see what they are doing to draw students.) The number of good seniors applying for NSF graduate fellowships in areas in pure or applied mathematics, other than in computer related areas, is smaller than the number of students recruited each year into the top ten mathematics departments plus about two leading programs in each of statistics, operations research, and classical applied mathematics. The demise of good undergraduate math majors today will see a decline in qualified graduate students in fields like statistics and operations research as well as mathematics in a few years. Many departments already have problems in obtaining enough qualified teaching assistants. Even when we do count undergraduate mathematics majors we are often "double counting" those with double majors in computer science (if separate from mathematics), economics, etc. On the other hand, we rarely consider the increasing number of students graduating in electrical engineering, operations research or modern industrial engineering who are essentially mathematics majors, some with an emphasis which could be called "modern applied mathematics". They are often well prepared to fill the jobs appropriate for math majors.

Some people dismiss the trends just cited with various other observations. Perhaps mathematics is just returning to its position in the 1950s. Whereas the current plunge in enrollments may in fact go right "through the floor" relative to the 1950s, and despite the great increased use of mathematics since then within the universities and beyond. Some would argue that such cycles of ups and downs in enrollments take place normally, and they seem willing to wait for the next spontaneous upswing, which still others feel is not about to occur. In contrast some are beginning to suggest that mathematicians had better "get with it" or risk a rapid decline to the point of being a department more like departments in classics or philosophy, or even a subunit within computer science. Trends in this direction could rapidly accelerate as other quantitative departments on campus decide to teach their own math "service" courses.

Clearly the problem of the "diminishing base" is a most critical one which deserves very serious attention from the highest levels of the mathematics community in the U.S., including the American Mathematical Society (AMS). It will not be sufficient to leave this to other associations of mathematicians or teachers, or to merely send a few delegates to national commissions. On the other hand, the main point for the purpose of this paper is to emphasize that major curriculum changes, such as significantly increasing the amount of discrete mathematics at the freshman-sophomore level, cannot overlook the impact this may have on attracting students to areas in the mathematical sciences and the ultimate health of the state of overall mathematics. It is not sufficient to merely lower some current upper level courses on finite mathematics, graph theory, or discrete structures, or else to provide courses primarily directed to one type of student (e.g., computer science majors) which are overloaded in one topic or approach (e.g., algorithms). Such courses must also contain additional "fun" parts of mathematics and develop the other mathematical skills deemed essential for mathematical sciences major. (See chapter 1 in the recent CUPM recommendations [5] for a discussion of these.) To avoid these broader objectives would risk aggravating the problem of the "diminishing base".

A proper solution for the problem of the diminishing base and the need for major curriculum changes will require cooperation across all levels and types of specialties in the mathematical sciences. It will not be sufficient for some subgroup (representing computer science, the core, or whatever) to push through their particular course changes while asking the larger community to cooperate and support them, or to at least remain "quiet". This point is important, since there currently exists some who would have us all cooperate to increase funding for core mathematics research and some special <u>parts</u> of applied mathematics, and seem to expect the excluded parts of "useful" mathematics to "cooperate".

4. MATHEMATICS USED TO BE FUN.

Any curriculum changes must pay serious attention to what draws people into mathematics and why they continue there after entering the field. There has been a tendency in the past two or three decades to strip the "fun" out of mathematics, to sterilize the subject to the point where all but the hardiest aspirants are not washed out, and to abstract and generalize prematurely. We tend to teach undergraduates the way we wish we were taught after we were well into or beyond graduate school, rather than the way we most enjoyed it as undergraduates. We too quickly rush past theories in two or three dimensions to do n-space. We often see freshmen with advanced placement in calculus who after one or two semesters of sophomore multivariable calculus with an A grade announce that they are switching out of mathematics because of diminished interest in this subject. Even when we introduce topics such as axiomatic systems the axioms are dictated and the students are given the "pie in the sky" advice that these are important in future courses. (E.g., the old myth about groups being necessary to take quantum mechanics while most members in such physics classes never really had such.) Whereas there are much more exciting models in which the axioms arise immediately in a natural way (e.g., social choice theory, voting models, apportionment, and fairness concepts). The historical context in which mathematical concepts arose are often neglected, as well as the personalities of the mathematicians so involved. Mathematics often appears to students as though it were an invariant over time and unrelated to its surrounding culture. And we often face the Hardy syndrome: The less useful, the better.

There are many quite different reasons why people enjoy mathematics and why they selected this subject for study or for career specialities. A selection of a dozen reasons appears in the following partial list. (1) Some people gained pride and confidence by solving high school mathematics problems faster than their classmates and decided to further pursue this natural talent or "road of least resistance". (2) Some consider mathematics as a fundamental subject and selected it somewhat because of the multiplicity of other disciplines and careers which remain open to those with a solid foundation in this most natural and basic field. (3) Many are impressed with the great power and ability of mathematics to solve problems in so many other diverse fields. E.g., the great success story of mathematics applied to theoretical physics in the earlier part of the 20th century. The relevance and utility of mathematics to modern science is truly amazing. (4) Even more exciting is the great reward from doing original mathematical modeling on one's own, i.e., of creating one's own model, solving it, interpreting the results, and arriving at new knowledge as well as verifying known facts. (5) Many are similarly excited by problem solving, finding new proofs of theorems, and creating new theorems as well. (6) The puzzle aspects and challenging game-like nature of much of mathematics appeals to many. (7) Some are truly fascinated with subjects like topology, or in applications of say geometry to areas such as relativity. (8) Others are "turned on" by the beautiful abstract structures within mathematics itself, such as those which arise in a modern algebra course, as well as more concrete geometric structures or other patterns. It is also remarkable how often beauty and utility come hand-in-hand in mathematics. (9) Some like their mathematics to be constructive or algorithmic in nature, i.e., they desire to "compute" and determine more concrete realizations of their theories. (10) Others are rewarded by becoming personally involved in logical precision, intellectual purity, and the rigorous arguments common to much of mathematics. (11) Whereas many are repeatedly surprised at the frequent connections which arise between mathematical structures and activities and experiences in everyday life. (12) The historical role of mathematics and its impact on modern culture is also an interesting topic to many people. Note that aspects of both pure and applied mathematics appear in this list.

It would be of interest to have individual mathematical scientists ponder the following questions, and to reply to them. When did you first become consciously aware of the fact that doing mathematics was "fun", i.e., an enjoyable activity, at least in part? When did you decide you might major in the mathematical sciences? When did it occur to you that you might make a career out of mathematics? Finally, can you recall what aspects of mathematics most appealed to you at these various times? It seems reasonable to conjecture that a broad variety of different answers would be given to these questions, especially the last one. It is surely a broad mix of personal attitudes and self satisfactions which lead different people into mathematics, and which appeal to or motivate students taking such subjects. In designing curriculum we must recall what was exciting about mathematics when we took it rather than later when, e.g., we taught it.

The above discussion about the variety of interesting aspects to mathematics should be considered very seriously by anyone attempting to implement major curriculum changes in the first two years of college. Such new courses are going up against tried and proven courses such as calculus and the basic physical sciences. They cannot be simply courses "created in committee" such as the discrete structures course which has been a failure in many cases. The new courses should use a "mixed strategy" in terms of appealing to more than one or two types of student interest. Yet they must rival the integrity and provide the overall cultural value of fundamental courses like calculus, without becoming a jumble of watered-down topics like the commonly taught finite math course for freshmen. There must be ample contents which are inherently interesting, useful to other fields, allow for a release of students' individual creativity, as well as serve as prerequisites for other courses.

It may well prove very important for various reasons to not let the new general finite math courses for lower division students become over specialized in one or another aspect of mathematics. A basic course or two devoted exclusively say to algorithms may not allow many students to develop other talents or interests. I have seen freshmen students gain much more satisfaction out of their <u>own</u> discovery of a basic theorem such as the max flow - min cut theorem than from successfully being able to compute a shortest path or maximum flow through a large network. There is also a historical precedent concerning the excessive concentration on algorithms.

In the century before Newton's time there was an extensive specialization on algorithms. However, this emphasis died out quickly after the discovery of calculus, and seems to have had less than major impact in the long run.

5. MATHEMATICS IS STILL ALIVE!

Another very serious problem in early college as well as precollege mathematics courses is the dreadful lack of any indication of the fact that mathematics is vital, dynamic, and currently relevant, with ongoing new discoveries occurring frequently. Mathematics is often viewed as "old stuff" which must be learned in order to pursue modern scientific subjects. Contrast this to the high school or first college course in physics or biology in which the student receives ample evidence of recent discoveries. Although most new famous mathematical (as well as scientific) discoveries cannot be explained thoroughly at elementary levels, there remains ample results to make this point at relatively lower levels. The problem is more one of selecting from the wealth of current results to include, and serious group discussion on this point is warranted. Above all, it is remiss in curriculum design to not undertake to demonstrate the "currency" of mathematics. This is particularly true in light of the lack of public understanding of the broad field of mathematics, and the substantial competition from such nonprerequisite media as television and its science programs.

6. RESISTANCE TO CHANGE.

It is clear that we can no longer afford to have the first two years of college mathematics dominated by calculus, nor the twelve years of precollege mathematics aimed primarily and nearly exclusively at preparing for calculus. There are needs and prerequisites for many students other than classical mathematics majors, physicists and engineers which must be introduced before the upper class years. It would also be worthwhile for those in the latter groups to also see some contemporary mathematics and its applications. Particularly since many such mathematics majors are dreadfully unaware of the broad nature of much of mathematics and the many available opportunities for such majors.

It is true that much of the resistance to curriculum change is due to conservative elements which resist dropping almost any well established course. Others are primarily interested in only pure mathematics and classical applied mathematics. Many also sincerely believe that calculus is the best course and is currently taught at the right time. On the other hand, there are those who argue rather persuasively that no discrete mathematics course with the real overall educational value and usefulness to rival calculus as a college entrance course has yet been designed or implemented. They could argue that most attempts to introduce discrete mathematics in the first two years of college have not been particularly successful. The typical elementary freshman finite mathematics course (sometimes done as math appreciation or in some other format) is a hodgepodge of little value as is rather clear from examining the many near "isomorphic" texts which publishers dictate in this subject. The discrete structures course has not been all that successful. Courses geared at only computer scientists or laden mostly with pure graph theory vocabulary and theorems, heavy on algorithms and short on real cultural value will not likely deserve a slot currently held for calculus or introductory science courses. In short, most existing elementary courses in discrete mathematics can hardly rival current calculus courses in integrity and quality at this time. It would appear that the advocates of these new courses should produce ample evidence that such courses of real value do or could exist. It is furthermore not sufficient to merely assert that several courses could profitably build upon elementary discrete mathematics, as is clearly the case for calculus, when so many students are taking such advanced courses without the earlier ones. (I might add that I am not an extremist like some mathematicians who criticize such fine books as the recent one by Davis and Hersh [1], The Mathematical Experience, because the nonmathematical readers might not get a clear picture of what mathematicians really do. I would prefer the public having this type of information rather than none which is the more likely alternative.) The main point is that we have quite a way to go before we have a worthy product in finite mathematics which can clearly be put up as a true rival of calculus, and in the meantime the burden of proof is on those advocating the discrete track.

It is essential that new finite mathematics courses be as culturally rich in some sense and of as high quality as calculus offerings. They must be rich in several of the "utility" and "fun" aspects of the subject without neglecting some depth component. They should contain ample individual problem solving and some experience at open ended modeling of real world situations. There is the chance of using highly motivating and realistic examples. Such material and experiments can be used in a highly interactive way so that students can experience personal involvement and self discovery in class as well as in doing homework assignments. It would be best if the courses were not exclusively lecture-exercise format, but involved students in more integral ways in self discovery and problem solving.

7. AVAILABLE MATERIALS.

One of the results of recent developments within mathematics has been the ample supply of new materials which can be used to design exciting new courses. One of my colleagues in mathematics and the physical sciences who teaches our junior math course, "Mathematics in the Real World", has argued that one cannot teach good modeling courses until the master's degree level. Whereas I would argue that recent developments in applied finite mathematics, operations research, quantitative social science, etc., have opened the way for exciting hands-on modeling experiences at the freshman and precollege levels. This is also true for many of the other "fun" aspects mentioned above. So it is most appropriate to involve many of these "fun" components in all elementary college courses in addition to merely covering the prerequisities needed by some subset of students, say those in computer science. It is beyond the scope of this paper to present sample lists of appropriate and precise topics for such new courses. Although there is room for more expository modules and books, there is already enough available for purposes of course design. There are several people at this Sloan Conference on the First Two Years of College Mathematics who have the experience and expertise to create such exciting lower level courses. Several here have developed such courses at the upper division level also. The major point is that there now exists materials and approaches to create truly superb finite mathematics courses at lower levels. Additional remarks by the author on materials appear in [3].

8. MULTIPLE OBJECTIVES.

The author believes that some undergraduate mathematics courses can successfully involve several approaches and objectives simultaneously. It is not necessary for a small department to have separate courses in linear programming, game theory (with mostly matrix games), discrete models, deterministic operations research, graph theory, combinatorial optimization, combinatorial problem solving, and networks. At the junior-senior level, much of this material could be integrated into about two courses which is about all most smaller departments can afford. Such a multipurpose course could cover several of the different "fun" topics, as well as represent both "pure" and "applied" type courses of each kind. (It is often overlooked that several of the subjects mentioned can be taught as pure or applied subjects, e.g., linear programming can be as pure as an algebra course. We tend to not make this distinction in finite mathematics, probably because of the shorter history, even though it is clearly present.) One could also argue that calculus already is a solid multipurpose course, but I will not go into detail on this point. A standard discrete mathematics course at the elementary level must similarly be a carefully integrated multipurpose course which does not sacrifice depth and integrity. This is particularly true if calculus is being delayed for a year or two. One could further assert that even greater overall efficiency can be obtained in this matter because, among others, it allows for appealing to a variety of students who find different aspects to be most interesting. There are of course limitations on what one can integrate. The failure of the CRICISAM project to integrate calculus with computers should be noted. Similar combinations of finite mathematics and introductory computer techniques could likewise be difficult to implement. Another point is the fact that it is fairly well known how advanced mathematics and science courses build upon our fairly standard calculus courses. As a more standard type of elementary finite mathematics course materializes, it will become imperative to not waste time in detailed repeating of such contents in more advanced courses, which is a serious inefficiency in most current programs.

9. AN ALTERNATE SOLUTION.

The author believes that a long range solution to squeezing finite mathematics into lower college levels will actually occur when it is pushed even lower, i.e., into the precollege curriculum. In light of the failure of the new math and the undesirability of merely going back to basics, it is obvious that we are ripe for embedding high school and junior high school math into a new matrix or overall framework. Several forms of finite mathematics should prove appropriate for this. Students could be exposed to problem solving in such a way as to lead them into graph theory or suchlike without having to force vocabulary or properties on them before they are already self-evident. A good place to test some of these possibilities without risking serious harm to the students involved would be in the programs for accelerating adolescents who show mathematical talent, e.g., at Johns Hopkins University. Such programs are often run by, and even have courses taught by, nonmathematicians who seem "hell bent" on rushing through the old classical curriculum to calculus and beyond.

Another way to find space may well be the opening up of an extra semester or year for mathematics in high school, without necessarily replacing advanced placement calculus. Recall that some twenty years ago we squeezed twelve years of precollege mathematics into eleven years for good students. I have talked with two experts in mathematical education in Israel who said that they may be about to squeeze the current eleven years into ten, at least for some accelerated students. This will open up another year, and they are pondering what should be done with it. Aspects of finite mathematics are being considered. (I was given the assignment of how to best describe the duality theorem of linear programming in two-dimensions for such high school juniors or seniors.) The teaching of finite mathematics in high school will of course require a great deal of teacher training.

10. CONCLUSIONS.

The case has been clearly made for the need of a high quality, one-year, solid and multipurpose course in discrete mathematics suitable for freshman and sophomore college students. It should be taken by most mathematical science students and not merely by those in computer science or areas such as the social and decision sciences. However, there are still pervasive arguments to the effect that the desired course is still in the formative stages, and that a good deal of additional course design, group discussion, and testing is still needed. Many such existing courses still have serious shortcomings. So the main problem is to get on with the creation, advertisement and implementation of superb and broad based discrete mathematics courses for freshmen which are without a doubt a true rival to challenge the established calculus sequence.

On the other hand, it is not clear that the case has yet been well made for a full two year sequence of discrete mathematics at the beginning college level. The topics frequently listed for the second year course often overlap substantially with other courses at different levels, and they appear to form less than a well integrated course. The burden of proof that there is real need for such a proposed second year course is still on those supporting such. The existence of only a one year course has many advantages, and appears as the best compromise at this time. It allows, e.g., for delaying calculus only one year. It also allows for two freshman-sophomore sequences consisting of 4 semesters of calculus and linear algebra, 2 semesters of finite mathematics, and 2 semesters of computer sciences. It further allows for some of the alternate "halfway house" measures discussed in Ralston's paper [4]. It also seems best to first have a well proven and accepted one year course in place before pushing too hard for the two year sequence. The latter may likewise become broadly implemented at some future time.

REFERENCES

1. Philip Davis and Reuben Hersh, The Mathematical Experience, Birkhäuser, Boston, 1980.

2. William F. Lucas, Growth and New Intuitions: Can We Met the Challenge? Chapter 6 in Mathematics Tomorrow, edited by Lynn Arthur Steen, Springer-Verlag, New York, 1981, pp. 55-69.

3. William F. Lucas, Operations Research: A Rich Source of Materials to Revitalize School Level Mathematics, Proceedings of the Fourth International Congress on Mathematical Education, 1980, to appear.

4. Anthony Ralston, Computer Science, Mathematics, and the Undergraduate Curriculum in Both, American Mathematical Monthly, 88, Aug.-Sept., 1981, pp. 472-485.

5. Recommendations for a General Mathematical Sciences Program, CUPM Report, edited by Alan C. Tucker, Math. Association of America, Washington, D.C., 1981.

DISCUSSION

Lucas: There are two major problems I wish to talk about:

- the administrative difficulties of making curriculum change within the constraints that growth as well as colleges and universities impose on departments; physics may not count for many students now but engineering is a major activity and their needs must be taken into account.
- the decline of mathematics enrollments generally; we must find ways to get people back into mathematics, and finite mathematics is one of the

possible ways; we need to make mathematics fun, both pure and applied.

Ralston: In your paper you propose that students interested in mathematics could take two mathematics courses a year during their first two years. But no one I talk to thinks you can do this and still satisfy the demands of a liberal arts curriculum.

Lucas: What I think you can push is one mathematics and one computer science course each semester.

Maurer: Do you think that algorithmic ideas are among those which could be used to add "fun" to mathematics courses?

Lucas: Yes and another area is modelling which I believe can be done at the freshman level.

Roberts: The comments about modelling and applications apply just as well to the standard calculus course as to discrete mathematics.

Anderson: You shouldn't be too pessimistic about mathematics enrollments. Social forces practically guarantee that the world is going to need more people trained in mathematics. There will be ebb and flow but not permanent decline.

Lucas: There is a counter argument. Cornell graduates about 40 mathematics majors a year of whom about half are joint majors with computer science. Of some 150 electrical engineering majors per year, many are for all practical purposes mathematics majors. Our operations research and industrial engineering students are also essentially mathematics majors. The point is that the mathematics majors aren't coming back; they're already going elsewhere to study mathematics.

THE MATHEMATICAL NEEDS OF STUDENTS IN THE PHYSICAL SCIENCES

Jack Lochhead

Cognitive Development Project
University of Massachusetts
Amherst, Massachusetts 01003

My task in this paper is to consider proposed changes in the mathematics curriculum from the point of view of a physics teacher. To do that I first need to distinguish between what students ought to be learning and what they do in fact learn. On the surface the present calculus curriculum is ideally suited to the needs of physics students; but a significant change seems in order when one examines the type of knowledge students are currently gaining. Thus the bulk of my paper will deal with certain problems in the calculus curriculum that I would hope will be effectively addressed by any new curriculum.

Before launching into a description of these problems let me first warn that much of the evidence I present may at first appear inconceivable. It was to those of us involved in discovering it; and that is why it has taken us years of research to reach even the limited understanding we now have. Second, I should also state that I am not claiming that students are not learning any of the mathematics we teach. On the contrary, they are learning a great deal. The problem lies not in the quantity of mathematics covered but in certain specific conceptual difficulties that severely limit the ways in which students can <u>use</u> the mathematical facts they do know.

I would have liked to start this paper with a detailed syllabus of the precise topics in mathematics essential to the study of physics. I am not going to do that, not because I do not believe such a list would be useful, but rather because I do not know how to construct one. For example, I am uncertain as to whether such basics as the Mean Value Theorem are really useful to students in their study of physics, or, for that matter, even calculus. The reason for my uncertainty is that I am so disturbed by some glaring deficiencies in the typical student's mathematical understanding that I question the value of almost all of their mathematical knowledge. The current curriculum is failing at a much more fundamental level than that expressed in the standard syllabi.

What Students Are Not Learning in Mathematics

Figure 1 illustrates some problems that many college students who have taken calculus still cannot solve. These results have been replicated at many different universities across the country and with a variety of different populations. Failure rates are amazingly stable. Student interviews and other types of tests have been used to show that the errors are not simply the result of carelessness or based on trivial mistakes. In fact it has proven to be quite difficult to teach students how to solve even the simpler examples (Rosnick and Clement, 1980). When Rosnick (1981) gave the problem in Figure 2 to students in the last week of their second semester of calculus he found that less than 60% could answer it correctly. More disturbing, 24% chose answers A4 and B1 (all students choosing B1 had picked A4).

Clearly these data indicate that many college students are not facile at reading or writing simple algebraic equations. But the problem goes deeper still. Students seem to lack any well defined notion of variable or of function. Figure 3 illustrates the difficulties a student had in talking about the meaning of one variable in a specific problem. Rosnick (1982) concluded that for most undergraduates,

WORD PROBLEMS MOST STUDENTS CANNOT SOLVE

Sample: freshman engineering students. Most were taking calculus at the time of the test. Very similar results have been obtained for calculus graduates.

	% Correct
Write an equation using the variables S and P to represent the following statement: "There are six times as many students as professors at this university." Use S for the number of students and P for the number of professors.	63
Write an equation using the variables C and S to represent the following statement: "At Mindy's restaurant, for every four people who order cheesecake, there are five people who order strudel." Let C represent the number of cheesecakes and S represent the number of strudels ordered.	27
Write an equation of the form P = for the price you should charge adults to ride your ferry boat in order to take in an average of D dollars on each trip. You have the following information: Your customers average 1 child for every 2 adults; children's tickets are half-price; your average load is L people (adults and children). Write your equation for P in terms of the variables D and L only.	2
Write a sentence in English that gives the same information as the following equation: A = 7S. A is the number of assemblers in a factory. S is the number of solderers in a factory.	29
Spies fly over the Norun Airplane Manufacturers and return with an aerial photograph of the new planes in the yard. * * * * * * * * * * * * * * red blue They are fairly certain that they have photographed a fair sample of one week's production. Write an equation using the letters R and B that describes the relationship between the number of red planes and the number of blue planes produced. The equation should allow you to calculate the number of red planes produced in a month.	32

Figure 1

S STANDS FOR PROFESSOR

Sample: students in a two semester calculus course for business and the social sciences.

Problem:

At this university, there are six times as many students as professors. This fact is represented by the equation $S = 6P$.

A) In this equation, what does the letter P stand for?

i) Professors
ii) Professor
iii) Number of professors
iv) None of the above
v) More than one of the above (if so, indicate which ones)
vi) Don't know

B) What does the letter S stand for?

i) Professor
ii) Student
iii) Students
iv) Number of students
v) None of the above
vi) More than one of the above (if so, indicate which ones)
vii) Don't know

Figure 2

THE VARIABLE MEANING OF VARIABLE

Problem:

I went to the store and bought the same number of books as records. Books cost two dollars each and records cost six dollars each. I spent $40 altogether.

Assuming that the equation 2B + 6R = 40 is correct, what is wrong, if anything, with the following reasoning. Be as detailed as possible.

$$2B + 6R = 40$$

Since B = R, I can write

$$2B + 6B = 40$$
$$8B = 40$$

This last equation says 8 books is equal to $40. So one book costs $5.

Excerpts from a student interview:

Student: ...Well, B is the number of books, but if-; more importantly-- in terms of figuring out how much they cost--B is-is a price, which is 2 dollars.

Interviewer: So you say B is the number of books and-

Student: at 2 dollars.

Interviewer: at 2 dollars.

Student: Well, no-; B-B is...B is the-is the um, the variable that equals the books at 2 dollars...

At other times during the half hour interview the student said:

1....Well, B is one book because it-it; B has something to do with the price of the book...

2....B equals the books...

3....B equals 5 times the price of that book-of the books which is $2.00...

4....Um...(11 sec)...No, I think B is equal to 1...but, um...it--I think you're referring to it right here...where you could say B is equal to 5...

Figure 3

variables do not have well defined referrents and instead represent an undifferentiated conglomerate of meanings. Based on this type of conceptual foundation, building a concept of function is at best problematical.

How is it possible for students who have such a poor command of mathematical language to advance through the curriculum? In fact it is quite easy. These students do have certain well developed skills:

- They can manipulate algebraic expressions (see Figure 4).
- Given numerical values for constants and variables in an expression, they can correctly plug these in and evaluate them (however, when one or more quantity is not given some students cannot progress).
- The better students can solve simultaneous equations and combine two equations to derive a third.
- Those who have had calculus can differentiate a variety of functions and may also be able to integrate or at least use a table of integrals.

Once a problem is expressed in algebraic terms students have little difficulty with it. Fortunately (or rather <u>most</u> unfortunately) textbook problems usually start with the formula. For example,

> An analysis of projectile motion shows the range to be $R = V_o^2/g \sin 2\theta$. What is the initial speed V_o of a projectile fired at an angle $\theta = 45$ such that its range is 800 meters?

One might think that a physics question such as the following demands more than symbol manipulation.

> A 4kg block slides down a frictionless ramp dropping 4 meters. Find its velocity at the bottom of the ramp.

But problems such as these can be solved in the following way. Make a list of the quantities given or asked for by the question. Choose a set of formulas, preferably no more than one or two, that includes all of these quantities. Make the appropriate substitutions and out comes the answer. The set of formulas used in a physics course is small enough to make this strategy highly reliable and efficient.

ALGEBRA PROBLEMS MOST STUDENTS CAN SOLVE

Sample: same students as in Figure 1

	% Correct
1. Solve for x: $5x = 50$	99
2. Solve for x: $6/4 = 30/x$	95
3. Solve for x in terms of a: $9a = 10x$	91
4. There are 8 times as many men as women at a particular school. 50 women go to the school. How many men go to the school?	94
5. Jones sometimes goes to visit his friend Lubhoft, driving 60 miles and using 3 gallons of gas. When he visits his friend Schwartz he drives 90 miles and uses ? gallons of gas. (Assume the same driving conditions in both cases).	93
6. At a Red Sox game there are 3 hot dog sellers for every 2 Coke sellers. There are 40 Coke sellers in all. How many hot dog sellers are there at this game?	93
7. Solve for y, given $y = \frac{x^2 + 3}{x}$ and $x = 9$	79
8. Solve for x: $\frac{(x^2 - x)\ (2x + 6)}{2x} = 0$	71
9. Solve for x: $\frac{13x - 7}{2} = 6x$	76

Figure 4

When I was in high school I wrote a computer program to do my physics homework. The program had a list of about 20 equations. When given a problem it would plug in the known quantities. Then it would try to do the calculations associated with all of the equations in its repertory. Of course only a few had sufficient information. It would then look to see if the quantity requested in the question had been assigned a value. If it had, that would be printed; if not, the program cycled back solving all of its equations again. Thus this system was capable of making complex substitutions, solving simultaneous equations, and carrying out long derivations? What I did not realize at the time was how accurately I had modeled the typical student's (and likely my own) grasp of physics.

One might argue that the lack of algebraic knowledge does not necessarily imply a poor grasp of the mathematical relationship described in formulas. Functions are perhaps more vividly described by graphs than by equations. Unfortunately our students do not seem to read graphs any better than they read equations. Figure 5 contains a list of questions about graphs together with some data on the performance of students who had taken first semester college calculus.

It has been my experience that most students can be taught to read graphs and to understand how they convey the nature of a relationship between two variables. I have had less success in getting them to reach an equivalent level of understanding for the algebraic expression of a function. But neither of these objectives are being effectively accomplished by the current curriculum at any point before or after calculus. When, if ever, (see Lochhead, 1980), they learn these ideas is, right now, a mystery.

GRAPHICAL DIFFICULTIES

% Correct

1. The graph shown is data for the location (x) of a car as a function of time (t). The location is given in meters and the time in seconds. — 22

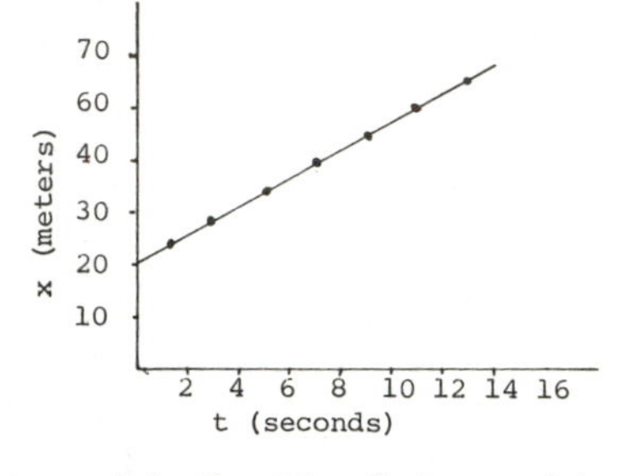

Given that speed is the rate of change of location, what is the speed of the car at t = 8 seconds?

2. A coin is tossed from point A straight up into the air and caught at point E.

What is the shape of a speed vs. time graph and an acceleration vs. time graph for the coin while it is in the air? Ignore friction.

C
B D
A E

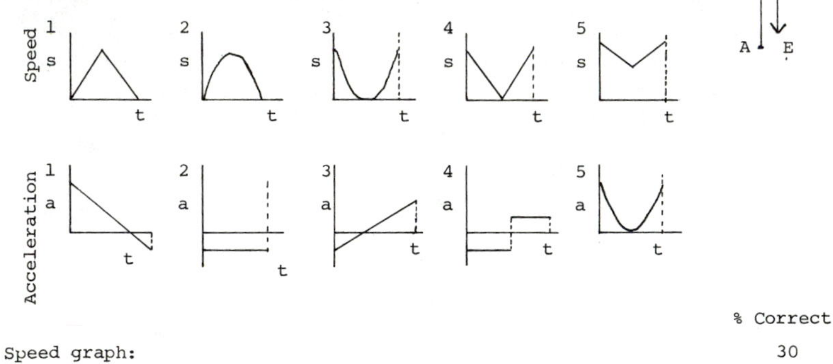

	% Correct
Speed graph:	30
Acceleration graph:	11

Figure 5

What Students Are Not Learning in Physics

Research into student conceptions of physics reveals that rather bizarre contradictions are commonplace. For example, students who have no difficulty solving problems using the equation $f = ma$ nonetheless seem to believe that force determines velocity (in essence $f = kv$). This type of thing is amazingly common even after two semesters of college physics (see Figures 6a and 6b and Clement, 1981, 1982). Students seem quite capable of memorizing and using the equations of physics without being aware of their content or understanding what relationships they describe. This type of learning might be compared to the knowledge held by a programmed hand calculator, only students are a good deal more difficult to program. In my opinion, this level of performance is not at all satisfactory. To improve the state of physics instruction physicists will have to place more emphasis on the qualitative aspects of physics. But they also will have to depend on mathematics instructors becoming more successful in teaching students how to employ mathematics as a language to represent relationships. It should be obvious to every entering physics student that if $f = m \frac{d^2x}{dt^2}$ then (in general) $f \neq k \frac{dx}{dt}$. I believe that one reason this is not evident to today's students is that for them equations are only rules for calculating and convey no useful information about relationships. Thus they make no attempt to relate equations to the rest of their knowledge. Any change in the mathematics curriculum that would effectively address this issue should be welcomed by physicists.

What is Missing?

There seem to be at least two fundamental causes of student difficulties with variables and functions. One is that students have not learned to be exceedingly precise and consistent in assigning a referent to a symbol. The second is that they tend to read algebraic equations as passive descriptions rather than as prescriptions for action. The equation $6S = P$ is seen as an exemplar case of 6 students and one professor rather than as an order to do something to the given number of students or a given number of professors (whatever that number may happen to be).

PHYSICS PROBLEMS MOST PHYSICS STUDENTS CANNOT SOLVE

1. A coin is tossed from point A straight up into the air and caught at point E. On the dot to the left of the drawing draw one or more arrows showing the direction of each force acting on the coin when it is at point B. (Draw longer arrows for larger forces).

   Typical Incorrect Answer: While the coin is on the way up, the "force from your hand", F_h, pushes up on the coin. On the way up it must be greater than F_g, otherwise the coin would be moving down.

2a) A rocket is moving along sideways in deep space, with its engine off, from point A to point B. It is not near any planets or other outside forces. Its engine is fired at point B and left on for two seconds while the rocket travels from point B to some point C. Draw in the shape of the path from B to C. (Show your best guess for this problem even if you are unsure of the answer).

b) Show the path from point C after the engine is turned off on the same drawing.

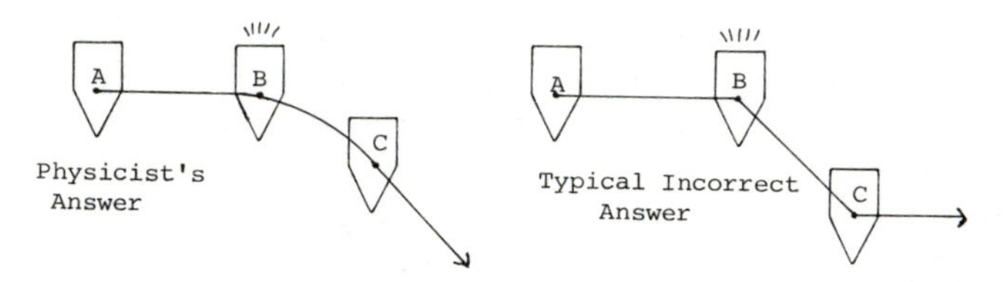

Figure 6a

	% Correct Before Course	% Correct After Course
COIN PROBLEM	12% N= 34	28% N= 43
ROCKET PROBLEM Part a)	11% N= 150	23% N= 43
Part b)	38% N= 150	72% N= 43

Figure 6b

When we first encountered the equation translation problem we hypothesized that if the students' difficulty were associated with a failure to grasp the dynamic operative nature of algebraic expressions then they should have less difficulty in a clearly operative domain such as a computer language. This is in fact the case (Soloway, Lochhead, and Clement, 1982). Even students who have had little experience with computer programming make fewer errors reading equations within programs than similar equations presented as algebra. There is some evidence that students with several semesters of programming make fewer errors in reading or writing algebraic equations than students who have just studied calculus. Thus computer programming appears to be a useful tool for teaching mathematics as a language.

There are several reasons why one might expect computer programming to be a better medium for teaching mathematics than algebra. Many of these have been described by Seymour Papert (1980). Two that are critical are the computer's penchant for doing precisely what it is asked to do, and the need (at least in lower level languages) of viewing each line of a program as a command for action. While from the mathematician's point of view algebra also has these properties it apparently does not in the student's mind. In fact, mathematical notation is deliberately ambiguous on these points, but the skilled user is rarely aware of the ambiguity because of contextual cues. For example, to the equation $y = mx$ one ought to assign a very different conceptual referent than to the equation $y_1 = mx_1$. (The fact that students do not know this is demonstrated by the difficulty students have with the equation for determining a line from a point and the slope, $y = m(x-x_0) + y_0$.) The cognitive referents of variables in fact change in slippery ways that are well understood (tacitly) by experts, but not necessarily by students (Rosnick, 1982).

The use of an even more sinister trick is responsible for much of the progress in modern mathematics. While it is important to be able to view functions dynamically, that same dynamical view is a serious nuisance in formal proof. Kaput (1979) expounds on this point using the example of limit. The original concept is dynamic and depends on terms such as approach, converge, and diverge. Yet formal mechanisms require a freezing of the action in the "for every .. there exists...". This ability to freeze action is extremely powerful and occurs in many situations. But it is also the source of much confusion on the part of students.

In my opinion, most students entering college have more need for the dynamic interpretation of mathematics than for the static. They need to understand what the subject is about before they can deal with formal (and relatively modern) proofs. Thus I feel we need to use terms and notations that capture the dynamic aspects (transformation, operator, mapping) and de-emphasize the static (equation). Computer languages seem to be an improvement over algebra, but may well involve conceptual costs that I am not yet aware of. The operator terminology, which is usually not used until Linear Algebra, may also be helpful. We have come to believe that in order to read an equation such as $S = 6 \times P$ one needs to view the $6 \times$ as an operator which acts to transform the number P into another number, S. This may seem an awkward way to handle a simple idea but our results suggest that there is no simpler way (Clement, Lochhead, and Monk, 1981).

What is Needed

I have claimed that the current curriculum is failing at a very basic level, and that if students are to grasp the fundamental ideas of mathematical language there needs to be greater emphasis on the dynamic nature of the process. I do not know whether or not discrete mathematics is likely to be an improvement. The notion of operator and transformation is inherent in much of the language of discrete mathematics but my guess is that it is no more central to the pedagogy than it is in calculus. One might think that the concept of a transformation between sets would be simpler for discrete sets than for continuous collections. I do not trust my own intuitions here and am even less sure when it comes to operations transforming functions. Perhaps continuous functions are easier to transform? In my opinion only careful research with students can resolve that sort of question. I do not believe that we now have the information to address the question of whether discrete mathematics is in some sense conceptually easier or more fundamental than continuous mathematics. To do that we will need to study students studying.

So far I have tried to show that what physics students most need from their mathematics preparation is a basic understanding of variables and functions and of how to express them in mathematical language. Contrary to popular belief this is a very tough criterion to meet. But specific content is not entirely irrelevant; students also ought to have the following:

- A feel for the operations of differentiation and integration. At the least it should be possible to connect equations such as:

  $$x = x_o + v_o t + 1/2at^2$$
  $$v = v_o + at$$
  $$a = \text{constant}$$

  This can be done graphically without the analytic techniques of differentiation and integration but I doubt most physicists would gladly give up completely on the analytic approach.

- An understanding of the concept of rate of change. For example, few physics students are clear on the distinction between velocity and acceleration. Again, calculus could be useful here; it may even be necessary, but it is not (in its present form) sufficient.

- Experience with vector addition, at least in two dimensions. This should include an understanding of the independence of orthogonal bases and how that relates to the process of decomposition into components. Trigonometric knowledge needs to be an integral part of this scheme; for many students trigonometry is a property of circles and has little to do with triangles.

- Finally, they should not lose the skill, which most now have, of being able to use equations to calculate the value of an unknown.

The above skills may be sufficient for the introductory student but they are not for those going on. Advanced skills must include:

- The ability to read and understand a function in several variables. This includes a feel for vector functions and vector operations such as the dot and cross product.
- The ability to read and write (set up) integrals in several dimensions. Most physicists would like their students to be able to evaluate such integrals with the usual techniques. I think numerical techniques (with calculators or computers) may be adequate (possibly preferable) but some serious questions need to be answered. In the usual second semester calculus course one learns substitution tricks which illustrate some very powerful aspects of mathematical thinking. This and other aspects of the course should not be dropped without careful consideration.

  Students' ability to set up integrals is currently hampered by a tendency to see integrals only as describing an area rather than also as a sum of elements. In integrals such as Work = $\int$ force x displacement = $\int F(x)dx$, the area concept is unwieldy.
- The following functions, their integrals and derivatives, should be familiar: Polynomials, at least up to the third degree; sin, cos, tan; exponentials and logs. There should be an understanding of how these functions describe growth and decay.
- The ability to coordinatize physical quantities, by representing them on a graph or conceptualizing them as elements of a multidimensional space.
- The ability to parametrize a function of one variable in terms of another. This includes transforming between coordinate systems and variable substitutions.
- The ability to read and write differential equations. I feel quite certain that existence proofs are a luxury. I think that some of the traditional curriculum should be replaced with numerical methods. Dartmouth College has had considerable success with this approach.

Conclusion

Most of the current calculus curriculum is essential to the study of physics. But much of what is covered in the early courses is not really learned or understood until very much later. It seems probable that "real calculus" could be delayed until the junior or senior year if certain aspects of the calculus were covered earlier. To do this would in some sense mean moving even further in the direction of teaching mathematics as a tool and leaving out the mathematical meat of the subject. Mathematicians are understandably reluctant to do that.

What I have tried to suggest in this paper is that the first step in teaching students to think mathematically needs to be a greater emphasis on using mathematics as a language for describing relationships. Without that basis there is really little point in developing a concept of proof. By mixing what I call qualitative or descriptive calculus with topics from discrete mathematics it may be possible to design a mathematics curriculum to satisfy everyone. Emphasis on the reading and writing of calculus expressions would satisfy the early needs of physicists and build a firm conceptual base for later math courses. At the same time, we could begin to develop students' appreciation for mathematical proof and rigor through studying subject areas that are less formidable than the calculus.

Acknowledgement

John Clement, Peter Rosnick, and Van Bluemel provided much of the data used in this paper. I thank them and Robert Gray, Ronald Narode, and Melvin Steinberg for their advice. Some of the work reported on in this paper was supported by NSF RISE grant 80-16567.

REFERENCES

Clement, J., "Students' Preconceptions in Introductory Mechanics", The American Journal of Physics, Vol 50, No.1, January, 1982.

Clement, J., "Solving Problems with Formulas: Some Limitations", Engineering Education, Vol 72, No.2, November, 1981.

Clement, J., Lochhead, J., and Monk, G., "Translation Difficulties in Learning Mathematics", American Mathematical Monthly, Vol 88, No.4, April, 1981.

Kaput, J., "Mathematics and Learning: Roots of Epistemological Status", in Cognitive Process Instruction, J. Lochhead and J. Clement (Eds), Philadelphia: Franklin Institute Press, 1979.

Lochhead, J., "Faculty Interpretations of Simple Algebraic Statements: The Professor's Side of the Equation", Journal of Mathematical Behavior, Vol 3, No.1, Autumn, 1980.

Papert, S., Mindstorms: Children, Computers, and Powerful Ideas, New York: Basic Books, Inc., 1980.

Rosnick, P., "The Use of Letters in Precalculus Algebra", unpublished Dissertation, School of Education, University of Massachusetts, Amherst, May, 1982.

Rosnick, P., "Some Misconceptions Concerning the Concept of Variable", The Mathematics Teacher, Vol 74, No.6, September, 1981.

Rosnick, P., and Clement, J., "Learning Without Understanding: The Effect of Tutoring Strategies on Algebra Misconceptions", Journal of Mathematical Behavior, Vol 3, No.1, Autumn, 1980.

Soloway, E., Lochhead, J., and Clement, J., "Does Computer Programming Enhance Problem Solving Ability?", in Computer Literacy: Issues and Directions for 1985, R. Seidel, R. Anderson, B. Hunter (Eds), proceedings of the Conference on National Computer Literacy Goals for 1985, New York: Academic Press, Inc., 1982.

DISCUSSION

Lochhead: The mathematics needed to understand the first two semesters of college physics is:

- simple differentiation and integration
- a feel for what differential equations are and for solutions in certain situations
- basic vector analysis in two and three dimensions

More generally, I think physicists would be sympathetic to changes in the mathematics curriculum which would orient it more in the direction of computers.

Greber: It needs to be noted that at many institutions physics is delayed until the start of the second year. And this allows much more flexibility in the mathematics curriculum.

Lochhead: We do that at the University of Massachusetts. Several people in our physics department believe that students should have at least a semester of college mathematics before they take physics.

Maurer: Physicists can also do a lot in terms of understanding of continuous processes using computers and without analytic mathematics. This seems to give mathematicians a fair amount of leeway in what they do with their introductory course.

Lochhead: I think that is precisely correct, particularly because front line physics these days consists of so much use of computer techniques.

ENGINEERING NEEDS AND THE COLLEGE MATHEMATICS CORE

Isaac Greber
Department of Mechanical and Aerospace Engineering
CASE WESTERN RESERVE UNIVERSITY
Cleveland, Ohio 44106/USA

Considering the variety of mathematical needs of engineering students, it is remarkable that essentially a common core of mathematics courses required of engineering students in the first two years of college has existed for so many years. This core has consisted of differential and integral calculus, the elements of analytic geometry needed for the calculus, some aspects of multivariable and vector calculus, and ordinary differential equations. The validity of this core as an appropriate background for current needs of engineers is being increasingly questioned, spurred to some extent by the burgeoning variety of needs of engineers but more importantly by the rapidly increasing use of digital computers in virtually all phases of engineering analysis and design. I will discuss some features of the changes of the core in response to technological developmnets and digital computation. My discussion will be limited mostly to the programs I know personally, and my viewpoints will probably be conservative. I will generally not distinguish between whether the courses are taken in high school or in college.

The traditional mathematics core is based on certain topics in the mathematics of continua. In a broad sense the core is directed toward the understanding and solution of differential equations. The background includes the understanding of limits and the concept of a function. Engineers of all kinds have traditionally regarded the ability to derive, solve, and understand differential equations as central to their analytical tools.

It is interesting that some important topics in continuum mathematics have traditionally been left out or only lightly emphasized in the mathematics core. These include ideas of complex variables (despite the fact that the Laplace transformation is frequently taught in the differential equations courses), integral equations, and the calculus of variations. Even before the widespread use of digital computers, engineers regarded some topics in non-continuum mathematics as fundamental to their work, but did not include them in the core. For most engineers these topics include probability and statistics, and for electrical engineers they include Boolean algebra. Engineering curricula tend to include topics in probability and statistics within the engineering courses, typically in courses on experimental methods, and electrical engineers include Boolean algebra in their electronic circuits courses. There is a more general import of the last remarks: much of the engineering student's education in mathematics takes place in engineering courses, and the perceived needs vary from curriculum to curriculum. The mathematics core represents both the topics that are regarded as "fundamental" and the ones that the students are commonly expected to know when taking other courses or solving engineering problems.

In response to the growing importance of digital computation, the mathematics core has typically been added to but not changed. For example, at Case Institute of Technology of Case Western Reserve University, in addition to the traditional core, all students are required to take a course in computer programming and a course in numerical methods. The course in computer programming centers around particular languages; currently Pascal is taught in some detail, and a small part of the course introduces Fortran. The numerical methods course includes integration, finding roots of functions, solving systems of linear equations, least squares fits and other curve fitting. Currently the computer language used in this course is APL. The major change in the courses that have been retained is the incorporation of some topics on matrices, including the definition of eigenvalues and eigenvectors, in the differential equations course.

In the late 1960's and early 1970's the freshman calculus course at Case was augmented by a "Mathematics Laboratory". This "laboratory" was intended to use numerical methods to emphasize some of the ideas of the calculus. Although practically speaking it covered some of the same topics as the current Numerical Methods course, its spirit was the connection between numerical techniques and the continuum ideas of the calculus.

One impact of computers is to provide a powerful tool for digitally handling problems that have been formulated in a continuum way. The typical example is the numerical solution of differential equations. Although discrete formulations of what are now regarded as continuum mechanics are likely to become more common, continuum formulations will remain vital because in some fields they are natural (e.g. fluid mechanics) and in some fields continuum modeling is exceedingly instructive. Within the framework of continuum mathematics, engineering and other students clearly need to learn programming (almost any language will do) and some numerical methods of dealing with functions. It is useful to realize that an increasing fraction of engineering students comes to college already knowing some computer programming, and we may soon find that college courses that emphasize programming per se will become merely remedial. Most numerical methods of solving differential and integral equations lead to solving systems of algebraic equations, so the need for studying linear algebra is clear, and is rapidly being incorporated into the mathematics core. More subtly, the use of numerical procedures requires greater sensitivity on the part of engineers to questions of existence and uniqueness; it is easy numerically to calculate one solution for a while, then unknowingly switch to another. Although most engineers will not construct existence or uniquesness proofs, engineering students should be made more aware of the importance of proper mathematical formulations of a problem than has been traditional.

The computer allows some topics in continuum mathematics to be de-emphasized or eliminated. The fact that a derivative of a function is zero at an extremum is an interesting piece of information, but is no longer the operational tool for finding the extremum. Similarly, much less attention needs to be given to techniques of performing integrations analytically. But the use of numerical techniques for evaluating continua (e.g. numerical integration) requires an interweaving of numerical and analytical ideas rather than separate courses.

A deeper re-examination of the mathematics curriculum stems from the fact that many engineers deal with problems that are "naturally" handled discretely (e.g. the trade interaction of a number of industries), that many physical situations that we tend to describe by differential equations can be described as statements about discrete systems (e.g. traffic on a highway can be thought of as a continuous flow or as a movement of a set of discrete cars), and that the computer allows us to handle discrete systems directly. In fact, we frequently reason in a circle; we start with a discrete formulation, pass to a limit to obtain a differential or integral equation, then obtain a discrete approximation to our continuum formulation. In some engineering fields the discrete system is more in the forefront than the continuous system, and the continuum is simply a way of modeling. Additionally, even where continuum modeling seems appropriate, the use of computational techniques may make it more useful to formulate a problem as an integral equation or a variational principle than as a differential equation.

Unfortunately, the answer to the needs of engineers isn't as simple as choosing discrete on continuum mathematics, or even changing the emphasis so that the primary emphasis is on discrete mathematics. The elements of continuum calculus and differential equations will continue to be a fundamental need of engineering students. They will in addition have to become comfortable with discrete mathematics not only as a way of approximating continua but as a way of dealing with problems from the beginning. To be effective, this attitude should be prevalent both in the mathematics and in the engineering courses. Similarly, if we are to keep the mathematics program from expanding too much and if discrete formulations are to be regarded as appropriate calculational tools where up to now we have taught analytical ones, then discrete mathematics should be incorporated into the continuum mathematics courses, in a manner which interweaves the two.

The admittedly major topics that we currently leave out of the traditional core, as a conscious choice, emphasizes our recognition that all needs cannot be met, and that not all topics are equally central to the variety of engineering specialties. In our search for curricula in keeping with development in computational power, and also in other new mathematics, we should keep in mind that we cannot do everything, and that we certainly must continue, as at present, to develop mathematical knowledge and maturity well beyond the accepted core, both in mathematics and engineering courses.

I've used the word "engineering" many times in this discussion. However, little that I have said is especially restricted to engineering, and I haven't cited many examples. A hallmark of engineering is variety, and the examples are exceedingly diverse. The core is something to build on, and if favorite topics are not included but mathematical maturity is developed, then the core is reasonable.

DISCUSSION

Greber: A point in my paper which needs emphasis is that you can't simply add courses in discrete mathematics because something has got to give. Thus, we have to try to get discrete topics interwoven with the present core which has esthetic advantages also. By the way, I agree completely that courses in mathematics are going to change whether we say so or not.

Universities have typically been lagging behind industry in using computers in the design process. I think engineers are starting to realize that finite mathematics has to become one of their mathematical tools. But in order for people who are used to continuous mathematics to accept the introduction of finite mathematics into the curriculum, it is important that computational tools for dealing with continuous phenomena be introduced early.

Pollak: At Bell Laboratories we have a great variety of engineers and generally they have a strong need for discrete mathematics. But is that peculiar to our place?

Greber: No, that is typical of electrical engineers and systems engineers, at least.

Maurer: It seems to be true in any field of study that you use the mathematical methods you have learned. Thus, if we put more discrete mathematics in the curriculum, engineers will find it useful, perhaps in surprising ways. Another point is that not just in engineering, but generally students will not learn all the mathematics they need from mathematics courses. I view this in a positive light because it means we don't have to cram in everything we think might be valuable to somebody.

Young: It seems to me the argument for the engineers is not whether a certain amount of discrete mathematics is going to help directly the civil engineer or mechanical engineer but rather that more discrete mathematics will mean better use of computers and, therefore, better engineers.

The First Two Years of Mathematics at the University as it relates to the mathematical needs of students in the social sciences

Robert Z Norman
Dartmouth College

The task of preparing a common set of courses in mathematics for students intending to major in one of the social sciences differs significantly from the same task for prospective physical science, mathematics, or engineering majors. For one thing, there is no generally accepted body of mathematics that every social science student is expected to know. In fact, a great many social science majors take as little mathematics as they can, usually a modicum of statistics. And few social science departments require much mathematics from their students. Besides, many of the faculty teaching in the social sciences have little mathematical training and use almost no mathematics; they are not apt to encourage their students to study mathematics. Many of these students and quite a few of the faculty show a strong fear of mathematics.

Another significant difference between majors in the physical and the social sciences is the time at which they choose their major topic. Majors in the physical sciences usually know on entrance that they expect to major in one of the physical sciences. A large portion of those who major in the social sciences decide to do so after the middle of their sophomore year. As a result, their choice of mathematics courses (or lack thereof!) in their first two years is often made without considering its relation to their particular major. (Economics students are somewhat of an exception to this general assertion.) (See also the paper by Gail Young in this book.)

Students who eventually major in one of the social sciences, if asked on arrival at the university for their intended major, would differ greatly in their responses. A significant number initially intend to major in mathematics or in one of the physical sciences or engineering. Some plan to major in the biological sciences or are pre-med students. Some intend to major in one of the social sciences. The rest, a large number, are headed for humanities, or simply "don't know." The first group begins its mathematical studies with courses for physical science students, and often takes four mathematics courses in the first two years. The second group typically takes a year of mathematics. The rest may take none at all, though increasingly many of the students with quite uncertain majors are taking some mathematics.

Another difference between the physical and the social science students is their interest in the calculus. For the physical sciences one finds immediate applications of the ideas of differentiation, integration, and differential equations. However, the dominance of these applications tends to decrease the interest of social science students. Social science students find applications come more easily in other areas of mathematics -- graphs, combinatorics, probability, linear algebra, and statistics. In reality, both types of mathematics -- calculus and the other areas -- have applications in both the social and physical sciences.

Particularly in view of the fact that social science students tend to decide on a major relatively late in their college careers, it is important to try to design a mathematics program for the first two years that does not force a student to make an irrevocable (or nearly so) decision at an early date, and that offers recognizable applications at an early point.

What courses do they need? I asked a number of social scientists what mathematics preparation they believe a social science major needs. I chose social scientists who use mathematics but who themselves are not mathematical social scientists. I told them my purpose was not to find the needs of budding mathematical social scientists, but rather the needs of a solid social scientist, and asked in such a way that the question of whether the curriculum allowed enough time was not of consideration. The topics, more or less in the order of the frequency and emphasis mentioned, are: Probability, manipulative algebra, computing (the ability to write a good program for simulation and for analysis of data), statistics (enough understanding to know what an appropriate statistical method is for the problem at hand), calculus and differential equations, combinatorics, linear algebra, sets and relations. Special topics included linear programming, game theory, geometry, and Fourier series. Calculus and differential equations was mentioned first by many, and very late by others. On further questioning of some of my later contacts, it seemed clear to me that respondents sometimes de-emphasized calculus and statistics because they knew any mathematican would automatically include calculus in any curriculum, and some of them thought of the first courses in statistics as being offered outside the mathematics/statistics department. Rates of change and growth and decay processes are topics of interest in many fields. At the close of our conversation, several concluded with the observation that most important was that mathematicians should train students to think clearly and logically and to be able to handle routine manipulative algebra.

I asked three biological scientists the same questions, and except for more emphasis on calculus and differential equations, got essentially the same answers as I did from social scientists.

Some of the respondents mentioned topics that seem special to their areas. In ecomonics, eigenvalues and eigenvectors, linear programming, matrix inversion. In psychology and sociology, eigenvectors (for foundations of factor analysis), Markov chains, data analysis and simulation, geometry, Fourier series. In political science, game theory, especially of n-person games.

It seems fair to draw the conclusion that the curricular needs in the first two years of university mathematics for students in mathematics, physical sciences, engineering, social and biological sciences are not as different as one might have expected.

In summary, social science majors adopt their major relatively late. Their previous intentions, if formulated, may lead them to take little mathematics or as much as physical science students. The needs of social science students in the first two years are quite similar to those of physical sciences students. Considering the desirability of getting more potential social science majors to elect mathematics courses so that when they become social science majors they will be mathematically prepared, it seems that an appropriate first two years of college mathematics for social scientists should be not very different from that for mathematicians, physicists, or engineers. The courses should be designed to be more attractive to social science students, who may choose to take the courses in a different order. It is important to design the courses so that few if any students have to repeat material or backtrack if they change their major. (For large universities it may be desirable to run sections for social science students at a less demanding pace.)

Outline of courses.

Semester 1. In view of the fact that the concepts of rate of change and of growth and decay are so prevalent in so many areas of study, the first course for all students is a semester of calculus. The course has as its goal the first-order differential equations of growth and decay. It is not radically different from current first courses in calculus. However it begins by posing the question that will be answered at the end of the course -- how to handle growth and decay problems in many different disciplines. The course could begin with sets and relations, and include mathematical induction, or this could be delayed to the second semester. The course can postpone much of the standard applications of the integral calculus to the second course in calculus. It should do integration algorithmically through using sums.

Semester 2. Two possible courses that can be taken in either order. One of these is a course that might be entitled discrete mathematics. It is clear that potential social sciences students would benefit from studying graphs and combinatorics, linear algebra, probability, statistics, and sets and relations, in addition to calculus. These cannot all be included in a second course. This course should begin with a little sets and relations, unless (preferably) that material is already known. It should continue with graph theory and combinatorics (and induction if not already studied), and should conclude with some discrete probability (through independent trials processes and conditional probability, so as to prepare students for statistics) or some linear algebra, if possible through the fundamental homomorphism theorem for linear transformations, and/or an introduction to eigenvalues.

The other choice for semester 2 is a continuation of the calculus. This continuation would result in a course much like that in the CUPM report, Recommendations for a General Mathematical Sciences Program [1], and would differ from the more standard calculus in that it would have less, much less, technique of integration, more differential equations, more on modeling -- not just "word problems," but putting more attention on assumptions being made -- and a little multivariable calculus, enough to be able to handle partial differentiation. Numerical solutions to integration, differential equations, etc., including estimating error, would be handled through use of a digital computer, which would also be used to aid in the development of concepts in class.

Semester 3. The most likely candidate is the other of the options for Semester 2 not yet taken. Other possibilities include, for those who took the discrete mathematics course already, statistics, or a course in probability, linear algebra, or graphs and combinatorics.

Semester 4. In addition to courses suggested above, another option is a third course in calculus, having additional material on multivariable calculus, and including Fourier Series.

Students who take only two semesters of mathematics should, of course, take the discrete course in the second semester. If they take only one year of mathematics they will probably take statistics in their social science department.

What about computer programming? A course in structured computer programming can be designed to be taken at any time concurrently with any of these courses, but the two courses mentioned for semester 2 are particularly attractive for this purpose. I would like to see programming taught concurrently with each of these, with programming and the mathematical ideas interwoven.

What about statistics? All social science students should, of course, take a course in statistics. A good general course in statistics which includes both standard parametric and nonparametric statistics can be done in one semester if one has probability through independent trials processes and conditional probability in advance, and if one is willing to leave out many of the proofs. The computer should play a major role in the course so that the students don't spend an undue amount of time grinding out numbers. The CUPM report [1] is a good guide, though I would include a modicum of exploratory data analysis in the course -- particularly as it provides a nonparametric analog of the regression line and talks about using data to generate as well as to test hypotheses.

The plan of the first four semesters offers an ideal program for future mathematics majors, for they will see, in their second or third course, enough different topics in mathematics that they will have a considerably better idea of what mathematics is about, and hence will have better information with which to decide on a major.

Can the discrete mathematics course come first? Yes, of course, it can. Yet if it does it is likely to become a weaker course than it will be if it follows a semester of calculus. Partly for this reason and partly for greater initial uniformity I propose calculus first. There can easily be a course in finite, or discrete, mathematics for students expecting to take no more math (outside statistics and computing). Many social science departments will insist on teaching their own statistics courses. The Mathematics Department (and/or a department of statistics) should offer a statistics course that will not intimidate the average social scientist, but will offer a somewhat more sophisticated course than most social science departments offer.

In general this program is for the eventual social science curriculum for the more mathematically trained social science major. It is not intended that one or two years of mathematics are enough to prepare the more sophisticated mathematical social scientist.

It is not even assumed that the more mathematically inclined social scientist would stop at the first two years. Courses in linear optimization, game theory, a second course in statistics, and more discrete mathematics are clearly good additional mathematical preparation for work in almost any of the social sciences.

There is being proposed here a distinct shift to a greater emphasis on discrete mathematics at the expense of analysis. While this note is particularly aimed at the social science student, the change is, and should be, taking place for students in the physical sciences as well. As one of my colleagues said, these changes are being considered, appropriately, concurrently with a corresponding change in the industrial base of our society.

[1] Alan C. Tucker, ed., Recommendations for a General Mathematical Sciences Program, CUPM Report, Mathematical Association of America, Washington, 1981.

DISCUSSION

Norman: If you think engineers are a diverse group, just look at social scientists. Moreover, as noted in my paper, social science majors are, by and large, not people who know whether they're going to be social science majors when they enter college. Some, for example, who were going to be engineers or physicists, have taken a fair amount of mathematics. On the other hand, to complete a major in a social science area in most colleges and universities, you don't need any mathematics at all. But I think we should concentrate not on the bottom half who will always succeed in avoiding almost everything but basically on people who are going to take a fair amount of mathematics and who are going to use some of that in their work.

Social scientists broadly will list a lot of different mathematics which they find useful. But I still make the basic assumption that a curriculum for the first two years can be reasonably universal.

Barrett: To what extent do social science students go back as juniors and seniors and take the mathematics they missed before?

Norman: My feeling is that there are a great many people who come into the social sciences late who will go back and take some mathematics.

Roberts: Much more than with other disciplines, we must expect that social science students will not take the whole two-year curriculum. Therefore, we must design the first year of a two-year curriculum with this in mind.

MATHEMATICS IN BUSINESS AND MANAGEMENT

Stanley Zionts
Professor of Management Science and Systems
State University of New York at Buffalo

Until the mid 1950's, the role of mathematics (and statistics to a lesser extent) in Business and Management programs was minimal: some elementary business mathematics which included a smattering of the mathematics of finance using interest tables and some algebra, as well as some cookbook statistics. Beginning roughly at that time, with the infusion of funds from various sources, business schools entered the scientific age. Numerous mathematical techniques -- some new, some not so new -- were assembled to help solve various problems of management. A particular discipline called Management Science sprang up. It incorporated various techniques -- mathematical programming, linear algebra, network methods, queuing theory, stochastic processes, statistics, recursive relations, and computer simulation to attack various management problems. Drawing upon the above techniques, management science has as its philosophy the solving of a problem.

As mathematics began to be established in business schools, applications of mathematics to various management problems became prevalent. Among them were the application of quadratic programming to financial portfolio analysis and to the planning of production, inventories, and work force as well as the application of linear programming to advertising media selection. Queuing theory was used to analyze service facilities, such as restaurants and banks. Other applications include the transportation method of linear programming, the economic lot size formula trading off inventory and order/set-up costs, CPM-PERT networks in project management, the use of learning curves, exponentially-weighted moving averages as a means of forecasting, and gravity models for site selection.

Models that are used most in practice include linear programming and computer simulation. These models have been applied in industry for approximately twenty years. Other models are also in use, but not nearly as wide-spread a use as linear programming and simulation.

Business schools differ in the extent to which they employ quantitative techniques. Some business school programs painstakingly avoid any reference to concepts of calculus and awkwardly work around instances where such concepts would be useful. For example, they might illustrate a simple maximization by enumerating all possibilities or by graphing the appropriate function. At the other extreme, there are business or management programs that require the equivalent of two semesters' worth of calculus plus one or two semesters of statistics and in some cases more.

Certainly both extremes make substantial cases for their positions. Which approach is best continues to be debated. My own background (undergraduate engineering, some graduate mathematics, and a Ph.D in industrial administration) and experiences make me a staunch supporter of the mathematical end of the spectrum. Our program in Buffalo is not as mathematically oriented as I would like, but it is certainly on the more mathematical end of the spectrum. (Bear in mind, more mathematically oriented reader, that "more mathematical" in a business school context may not be very mathematical from the perspective of a mathematician.)

I would now like to describe our program at the State University of New York at Buffalo and describe what I see to be its strengths and weaknesses. In our undergraduate program, students take a two-semester sequence in calculus and linear algebra offered by the mathematics department. The course is about one and a half semesters of calculus with applications and a half semester of linear algebra, also with applications. A list of topics covered is included, excluding linear programming and trigonometric functions, in the appendix. They also take a two semester sequence in statistics and computing which covers BASIC programming, probability and statistics, and the use of the computer for statistical analysis. The first semester of that sequence is taught by the Department of Educational Psychology. Our school teaches the second semester. These courses are followed by a course titled Production and Operations Management, which includes a treatment of the production activities of a firm and an introduction to the philosophy and methodology of Management Science.

In this course students use packaged computer programs for solving problems; they also write computer programs in the solution of certain other problems. Subsequent undergraduate courses build on the above foundation. For example, economics makes use of the calculus (both differential and integral), certain functional areas (such as marketing and finance) use some of the models used in the above courses, and several courses use the computer skills.

In the graduate (Master of Business Administration) program, our students take a scaled down version of what we require of the undergraduates. For starters, no mathematical background is required for admission. Students who appear particularly weak in mathematics are advised but not required to take a noncredit undergraduate college algebra course. Entering students take two-and-a-half one-semester courses in the quantitative methods and computer areas. The first one-semester course includes a review of algebra and introduction to differential and integral calculus, matrix algebra, linear programming, and simulation, all with applications. (The mathematics of finance is covered as a part of the exponential function.) I have taught this ambitious course for several years. Although I have occasionally taken a ribbing or two at the annual student banquet (one year I received a student award for teaching all of integral calculus in one session!), I do believe the students' comprehension of the material covered is reasonably good for our purposes by the end of the semester. I have developed teaching notes for the course; a copy of the table of contents is found in the appendix to the paper*. The notes are also appropriate for an undergraduate course. Admittedly, students with a poor mathematical background have to work hard, but numerous students who entered the course with poor mathematical backgrounds have done extremely well in the course.

A second one semester course covers probability and statistics. It covers probability theory, inference, and descriptive statistics in considerable detail. It makes use of standard statistical computer packages and employs them in various analyses. This course emphasizes managerial applications of statistical methods, and is also quite ambitious in the amount of material covered.

An additional half-semester course provides an introduction to computers and computer programming and management information systems. The course begins with an overview of computers, their concepts and technology, introduces BASIC, FORTRAN, text editing, and the operating system used at SUNY Buffalo. The course attempts to create an awareness on the part of the student of the computer as a valuable problem-solving tool for management.

Entering students having sufficient background may waive one or more of the above courses.

Subsequent courses build on the above materials and apply them to problems in various areas of management -- accounting, economics, finance, human resources, and production. In some courses additional models that are desired are introduced.

We also have more advanced courses for people who have a more substantial interest in Management Science or Management Information Systems, or both. These courses vary in mathematical content, depending upon the objectives and content of the course, but prepare students for positions as analysts and problem solvers.

---

*Copies are available upon request from the author.

The above presentation is a statement of what constitute our required courses in the mathematics/computing area at the School of Management, State University of New York at Buffalo. All three courses are ambitious in their coverage. From these courses the students do achieve a certain level of knowledge in the material covered. By no means are they expert in all material covered. They do master certain techniques. Students completing our program gain from the above courses a modicum of mathematical maturity.

A conference such as this addresses what should be the content of an introductory mathematics program for a business/management program as well as for other areas. I would address this question in the following way: If we were to have additional time in such courses, to what subjects should they be devoted? Based on our experiences both at the undergraduate and graduate (M.B.A.) levels, I would devote more time to subjects already covered. Students at the undergraduate level particularly require more reinforcement of the material covered. M.B.A. students could also benefit from additional time devoted to the topics covered.

Much as I would like to add material to both programs, the prospects for so doing at this time are sufficiently remote as to say with virtual certainty that we cannot add material. Further, the current mix of material covered in our courses is just about right: I could not recommend any changes in content at this time.

To try to be a bit more positive about the objectives of the conference, and what is accomplished, I would like to encourage an evolutionary approach to the question of what should be taught in an undergraduate curriculum. Make some incremental changes from the current curricula, try it and evaluate it. Our programs already include some discrete mathematics, for example. To add more, something else must go, at least unless you can sell our programs additional credit hours of mathematics. Further, if it is proposed to replace calculus, then how do we handle what we use the concepts of calculus for? In the evolutionary process more interaction should be had between mathematicians, computer scientists, and the service community (engineering, sciences, management, and social sciences) whose students populate the undergraduate mathematics and computer courses.

Appendix

MATHEMATICAL MODELS FOR MANAGEMENT
(Working Title)
An Introductory Text

Stanley Zionts
Professor of Management Science and Systems
School of Management
State University of New York at Buffalo
Buffalo, NY 14214
716-831-2311

(Copies of chapters are available upon request from the author.)

The purpose of this book is to provide an introduction to mathematical models for management for M.B.A. students (and possibly undergraduates in management). The book is for a one-semester course beginning with basic mathematics (sets, functions, etc.) and progressing through differential calculus, integral calculus, matrix algebra, and linear programming. (At SUNYAB, a similar undergraduate in two semesters course is taught by the Mathematics Department.) Though rigorous, the book emphasizes application and is particularly oriented toward a mature student. The book is being written by the author for a course at SUNYAB, and is evolving over a period of years during which it is being class tested.

Tentative Table of Contents

II. Nonlinear Functions

A. Quadratic and Polynomial Functions and Their Applications
B. Exponential Functions
C. Logarithmic Functions
D. Applications of Exponential and Logarithmic Functions including The Mathematics of Finance
E. Trigonometric Functions: Their Definitions

III. Limits and Differential Calculus

A. Limits
B. Continuity
C. Derivatives: What They Mean and How to Compute Them
D. Applications: Maxima and Minima
E. Higher Order Derivatives*

IV. Functions of Two or More Variables

A. Partial Derivatives and Differentiation*
B. Applications of Partial Derivatives in Management*
C. Unconstrained Optimization and Applications in Management*
D. Constrained Optimization and Applications in Management*

V. Integral Calculus

A. Definite Integrals and Integration
B. Indefinite Integrals and Integration
C. The Technique of Integration
D. Multiple Integrals
E. Numerical Integration
F. Applications in Management

VI. Matrix Algebra

A. Vectors and Matrices
B. Matrix Operations
C. Special Matrices
D. The Transpose of a Matrix
E. The Inverse of a Square Matrix
F. Rank of a Matrix
G. Applications in Management

VII. Introduction to Management Science

A. An Overview*

B. Simultaneous Linear Equations*

C. Linear Programming*

1. Basic concepts introduced*

2. Solving the linear programming problem--the simplex method*

D. Duality*

E. Sensitivity Analysis*

F. Applications in Management*

G. Overview of Other Methods of Management Science*

DISCUSSION

Zionts: The questions management (or any other) faculty should ask are:

Why is the curriculum you are proposing better for us?

Why is it better for our students?

Curricula are always changing. Some sensible revision of the mathematics curriculum influenced by computers is certainly desirable. But we must keep the above questions in mind as well as the inevitable political and budgetary problems.

Ralston: How much of the calculus sometimes required of management students is actually used in later undergraduate courses?

Zionts: The skills are used in several courses but the modicum of mathematical maturity that is developed is very important.

Pollak: How about the statistics course taken by your students? Is it more data analysis oriented or is there a lot of formal statistics?

Zionts: The course is supposed to be data analysis and computer oriented but, depending upon who teaches this course, there can be some problems with too much formality.

*Drafts of these materials are not yet available. They should be available during the Fall, 1982 semester.

Mathematics Curriculum and the Needs of Computer Science

William L. Scherlis and Mary Shaw
Computer Science Department
Carnegie-Mellon University
Pittsburgh, PA 15213

Some Words about Computer Science

Computer science is concerned with the phenomena surrounding computers and computation; it embraces the study of algorithms, the representation and organization of information, the management of complexity, and the relationship between computers and their users. Computer science is like engineering in that it is largely a problem-solving discipline, concerned with the design and construction of systems. But the computer scientist, like the mathematician, must be able to make deliberate use of the intellectual tools of abstraction and of analysis and synthesis. The relationship between computer science and mathematics is very close and has been discussed at length in the literature. Two very interesting examinations of this relationship are [3] and [5].

Computer science is a mathematical discipline --- so much so that the boundary between computer science and mathematics is often quite hard to pin down. While both disciplines are concerned primarily with abstract structures, computer science is not simply a branch of mathematics. It relies on skills, attitudes, and techniques derived from mathematics, but it is concerned not so much with proofs and the existence of structures as it is with algorithms and the design and organization of structures. In this sense computer science is an engineering discipline. Like engineering, it is pragmatic and empirical and is concerned with the selection, evaluation, and comparison of designs for implementation. But in computer science this study is focused on the behavior of systems such as algorithms, computer organizations, and data representations --- that is, on abstract rather than on concrete systems.

This paper addresses the mathematical component of a good undergraduate computer science curriculum. It begins by describing the general nature of the mathematical needs of computer science undergraduates and then discusses some specific mathematical topics that are particularly helpful in computer science education. These mathematical topics include not only traditional mathematical subjects that can be taught in self-contained courses, such as discrete mathematics, but also certain mathematical *modes of thought* that pervade computer science thinking and that cannot be taught easily on their own. In the last sections we consider the impact of these needs on the curriculum.

Mathematical Aspects of Undergraduate Computer Science

There is a persistent misconception that computer science consists merely of writing computer programs and that, as a result, the education of a computer scientist consists merely of training in skills related to coding and debugging computer programs. On the contrary, the discipline embraces principles and techniques for the design, construction, and analysis of a wide variety of complex systems. Even programming, to be successful, requires the careful application of scientific principles.

Since the principles of computer science are largely mathematical, computer science curricula must necessarily rely on support from mathematics. The traditional mathematics and applied mathematics "service" curricula, steeped as they are in continuous mathematics, do not, however, provide adequate support for computer science. The demands of computer science on mathematics are in many respects quite different from the demands of traditional scientific or engineering disciplines. The most important difference is that, to a much greater extent than in other disciplines, *abstraction* is an essential tool of every computer scientist, not just of the theoretician. The computer scientist is not simply a user of mathematical *results*; he must use his mathematical tools in much the same way as a mathematician does.

A computer science undergraduate curriculum must attempt to develop in the student an appreciation of the power of abstraction and an ability to discover abstractions suitable to new situations. This ability is what mathematicians call *mathematical maturity* (see [13] for further discussion). Mathematical maturity will not be fostered if mathematics is taught to computer science students as a mere skill or as an unpleasant necessity.

Like other scientific and engineering disciplines, computer science must also teach certain specific attitudes, skills, and techniques from mathematics. In computer science, most of these come from *discrete mathematics* --- the mathematics dealing primarily with discrete objects. Discrete mathematics as an independent subject is a relatively new arrival, however, and present courses in this area often do not have the cohesion or intrinsic interest of the traditional calculus or algebra sequences. It is interesting, however, that many discrete mathematics courses use the notion of algorithm --- a concept from computer science --- as their unifying element [10, 12, 14].

Mathematical Modes of Thought Used by Computer Scientists

The most important contribution a mathematics curriculum can make to computer science is the one least likely to be encapsulated as an individual course: a deep appreciation of the modes of thought that characterize mathematics. We distinguish here two elements of mathematical thinking that are also crucial to computer science and speculate on how they might be integrated into a mathematics curriculum. These elements tend not to fall into identifiable courses, but are generally transmitted *culturally*, as part of the process of attaining that elusive quality of mathematical maturity. The two elements are the dual techniques of *abstraction and realization* and of *problem-solving*.

Abstraction and Realization

Computer scientists usually deal with situations that are too complicated to understand completely at one time. The chief tool for managing this complexity is *abstraction* --- a process of drawing away from detail or selectively ignoring structure. Conversely, complex real systems are built from abstract characterizations by the inverse process of *realization* or *representation* --- the selective introduction of underlying structure.

In mathematics, the deliberate use of abstraction is most noticeably manifest in the notion of *mathematical system*. The mathematical systems that are most useful to mathematicians, such as groups, fields, or categories, are those that best focus recurring problems. In computer science this kind of abstraction or encapsulation appears in many forms. Finite state automata, for example, permit study of control flow in programs without reference to variables or data.

Mathematics can be characterized by its search for gems of abstraction --- those abstractions that capture the essential qualities of a phenomenon and ignore the rest. Computer scientists carry on a similar search, but, because the structures they describe usually become manifest as real systems, they are concerned with the *performance* of systems as well as with their functional properties. Consequently, computer scientists find they are often fighting two sides of the same battle: Given a complex problem, they must develop abstractions that provide a way of managing the complexity, allowing for clear and effective reasoning about the problem. But they must also ensure that the representations or realizations that are hidden beneath their abstractions yield implementations with satisfactory performance.

The computer scientist who appreciates the variety of mathematical systems will be better able to evaluate structures and organizations for program and system design. A student who becomes comfortable thinking in terms of systems will be more likely to appreciate the full generality of the program or system structures he creates and less likely to think only in terms of the present specific application.

To strike the best balance between clarity and performance, the computer scientist needs a large and varied arsenal of abstraction and realization techniques. Some of these are rooted in conventional computer science and are therefore most appropriately taught in the context of computer science problems. Others, however, are best transmitted through a comprehensive study of mathematical reasoning.

One of the most powerful tools for abstraction is *language*. For example, programming languages are languages that allow the expression of algorithms without reference to particular realizations of algorithms in computer hardware. These languages also give us a way of describing data by means of its structure, not by its representation as "bits" in a computer memory. Like mathematical languages, computer languages are not designed in a purely *ad hoc* fashion; they are, rather, manifestations of carefully chosen lines of abstraction. If a computer science student is to appreciate the variety and universality of computer languages, he or she must have a mathematician's understanding of the nature and use of language. This includes, for example, understanding the nature of symbols and the essence of deduction --- carrying out worldly reasoning by means of symbol manipulation.

This discussion does not, alas, point to courses from "traditional" computer science curricula [2, 4] that will satisfy this need. (Indeed, the standard curriculum designs barely acknowledge the fact that exposure to mathematical reasoning is appropriate for computer science [9, 10].) There are courses in mathematics, however, that can foster the kind of understanding we seek. A good logic course, giving a kind of introspective view of mathematical reasoning, can be of great benefit to the computer scientist. Other mathematics courses, such as the analysis courses that are intended for mathematicians (as opposed to the ones intended for calculus "users"), can be of value simply because of the experience in mathematical definition and reasoning that the students obtain.

Problem-solving

Computer science is a problem-solving discipline, concerned with the development of cost-effective solutions (such as programs and machines) to computational problems. Computational problems do not in general have predictable structure and are almost always stated in abstract terms. As a consequence, the construction of programs (or even machine architectures) is analogous to the construction of mathematical proofs. While a proof (or program) has a well-defined structure, the process of obtaining it can be quite undisciplined, involving all sorts of peripheral and heuristic knowledge. Thus, the computer scientist, like the mathematician, must have command of a variety of problem-solving techniques, and must be able to apply them in a creative and yet disciplined fashion.

The designers of many graduate curricula in computer science have acknowledged the importance of abstract problem-solving and have incorporated problem-solving workshops based on such texts as [8] into their programs. We suggest that this need should be directly addressed in undergraduate curriculum design [15]. It is very important for students to be aware of the problem-solving process and of the general techniques that they can apply to it. Courses on these topics have been offered in engineering and computer science departments using texts such as [11, 16], but they could be equally appropriate in mathematics departments.

Discrete Mathematics

In addition to the ability to think like a mathematician, a computer scientist requires fluency in some specific areas of mathematics. These are the areas usually (collectively) called discrete mathematics, and they include such topics as elementary set theory and logic, abstract algebra, and combinatorics. Since this material is well-understood, an outline should suffice:

- *Elementary Set Theory and Logic.* It is important that the treatment of logic go beyond the usual manipulative knowledge of the propositional connectives and quantifiers. Students should have an appreciation of the central issues of mathematical logic and in particular of the role of language in mathematical definition and reasoning. This appreciation can be brought out both in the subject material *per se* and in the way it is presented.

- *Induction and Recursion.* These are recurring themes in computer science and should be covered in depth; induction underlies nearly all techniques for reasoning about the correctness and performance of programs.

- *Relations, Graphs, Orderings, and Functions.* This is a part of basic mathematical fluency. Without this knowledge, it is hard to understand even the most basic algorithms.

- *Abstract Algebra.* Algebraic structures recur in computer science, particularly in automata theory, complexity, software specification, and coding theory. A good introduction to algebra will develop in the student an understanding of the notion of mathematical system and will give him experience in using several of the more common ones.

- *Combinatorial Mathematics.* Analysis of algorithms requires a wide variety of mathematical skills; these are drawn mostly from combinatorial mathematics and from probability and statistics.

Although we have not as yet found a completely satisfactory text for discrete mathematics in computer science, the books [6, 12, 14] can serve as a starting point.

Continuous Mathematics

Although our primary emphasis here has been on the role of discrete mathematics in the computer science curriculum, we believe that continuous mathematics is also important to the education of a computer scientist. A mathematician's calculus course can serve as an excellent introduction to mathematical thinking. We will need to consider the question of when calculus should appear in the curriculum. For the purposes of computer science courses, discrete mathematics should appear as early as possible, preferably in the freshman year, but it has also been argued that calculus should precede discrete mathematics in the mathematics curriculum.

Some Remarks about Computer Science and Mathematics Curricula

As we noted above, the undergraduate computer science curriculum designs currently endorsed by major professional organizations have very weak mathematics requirements [2, 4]. Perhaps this is only a side-effect of the recent rapid growth of undergraduate computer science, but in any case it is widely viewed as a shortcoming. (See [9] and reactions to that article.) It is interesting to note that *earlier* computer science curriculum designs [1] contained much stronger mathematical requirements. Comparisons of the early and recent curricula are given in [9, 10].

With respect to the mathematics curriculum, we believe that support for the ideas and topics listed here would not cause major disruption to most mathematics curricula. The most significant change would be the addition of a freshman- or sophomore-level course in discrete mathematics. We believe that this course would be beneficial to students in other departments as well as to computer scientists. (The case for teaching elementary discrete mathematics to all students is presented by Ralston in [10].) Beyond that, most of the material we propose is fairly standard, though perhaps different in emphasis from in the traditional mathematics service courses. We should note here that our list should in no way be construed as complete; we mention topics only to provide an indication of the kind of material that is relevant.

Although much of the material computer scientists need is already provided in standard courses, we believe that both computer science and mathematics curricula would be strengthened by recasting some of those courses a bit. Teachers of mathematics can take advantage of their students' knowledge of computers by showing how classical techniques are realized in computational systems and, where appropriate, by drawing on the rich collection of practical examples supplied by computer science. Linear algebra and numerical analysis courses already do this, teaching computational techniques along with abstract definitions. Discrete mathematics, combinatorics, and graph theory courses also often make extensive use of programming exercises. These programming exercises give students an unusual "hands-on" way of experimenting with abstract structures. Moreover, Lochhead [7] argues that programming *per se* contributes to understanding mathematical ideas.

Conclusion

Computer science as a discipline has reached the point where there is enough intellectual substance for undergraduate degree programs to be meaningfully offered. Computer science courses are no longer simply programming "service" courses offered for the benefit of computer users; there is truly fundamental conceptual material to be imparted.

A successful undergraduate curriculum, in which basic principles are set forth and elucidated, can only come about after intensive self-examination in the field. Naturally enough, there is a certain lag between the time these principles first emerge and the time they can be effectively integrated into a curriculum, but we feel that there is now a consensus among computer science researchers and practitioners regarding the mathematical content of the field, as sketched in this paper. This consensus, unfortunately, does not extend to the methods for imparting the mathematical material; this remains one of the central challenges of computer science and mathematics curriculum design.

Acknowledgements

We thank Roy Ogawa and Dana Scott for their helpful comments on an earlier manuscript.

References

1. ACM Curriculum Committee on Computer Science. "Curriculum 68: Recommendations for Academic Programs in Computer Science." *Communications of the ACM 11*, 3 (March 1968), 151-197.

2. ACM Curriculum Committee on Computer Science. "Curriculum'78: Recommendations for the Undergraduate Program in Computer Science." *Communications of the ACM 22*, 3 (March 1979), 147-166.

3. Bruce W. Arden (ed.). *What Can Be Automated? The Computer Science and Engineering Research Study (COSERS).* MIT Press, 1981.

4. Education Committee (Model Curriculum Subcommittee) of the IEEE Computer Society. A Curriculum in Computer Science and Engineering Committee Report. IEEE Computer Society, November, 1976.

5. Donald E. Knuth. "Computer Science and Its Relation to Mathematics." *American Mathematical Monthly 81*, 4 (April 1974).

6. C.L. Liu. *Elements of Discrete Mathematics.* McGraw-Hill, 1977.

7. Jack Lochhead. Math for Physics. Proceedings of Sloan Foundation Conference/Workshop on Undergraduate Mathematics, Williams College 1982, 1982.

8. George Polya. *How to Solve It.* Princeton University Press, 1973.

9. Anthony Ralston and Mary Shaw. "Curriculum '78 -- Is Computer Science Really that Unmathematical?" *Communications of the ACM 23*, 2 (February 1980), 67-70.

10. Anthony Ralston. "Computer Science, Mathematics, and the Undergraduate Curricula in Both." *American Mathematical Monthly 88*, 7 (1981).

11. Moshe F. Rubinstein. *Patterns of Problem Solving.* Prentice-Hall, Inc., 1975.

12. D.F. Stanat and D.F. McAlister. *Discrete Mathematics in Computer Science.* Prentice-Hall, Inc., 1977.

13. Lynn Arthur Steen. Developing Mathematical Maturity. Proceedings of Sloan Foundation Conference/Workshop on Undergraduate Mathematics, Williams College 1982, 1982.

14. J.P. Tremblay and R.P. Manohar. *Discrete Mathematical Structures With Applications to Computer Science.* McGraw Hill, 1975.

15. D.T. Tuma and F. Reif. *Problem Solving and Education: Issues in Teaching and Research.* Lawrence Erlbaum Associates, 1980.

16. Wayne A. Wickelgren. *How to Solve Problems.* W.H. Freeman and Company, 1974.

DISCUSSION

Scherlis: First I want to emphasize three things:

1. Although computer science owes its existence to an artifact, the computer, it is not the study of the use of computers but rather the study of phenomena surrounding computing. Ten years from now the subject may be very different from what it is today.
2. Many computer science specialities, for example, numerical analysis and the theory of computation, were in existence before computers appeared. The arrival of computers, however, caused the connections between these areas to be made.
3. Computer science is certainly very close to mathematics. Both deal primarily with abstract structures but, while mathematicians are concerned with proof and the existence of structures, computer scientists are concerned with algorithms and the design and organization of structures. In this sense computer science is an engineering discipline. There is a pragmatic element in computer science that is not ordinarily a part of mathematical thinking. But computer science students do have to attain a fairly high level of mathematical maturity and have to be able to think like mathematics students.

The role of abstraction in computer science needs some elucidation. It is _the_ tool for managing complexity. But because computer scientists are also concerned with performance, they must sometimes compromise in their search for abstractions. Appropriate in this context is a remark of Jack Schwartz in 1968: "In this quest for simplification, mathematics stands to computer science as diamond mining to coal mining. The former is a search for gems. Although it may involve the preliminary handling of masses of raw material, it culminates in an exquisite item free of dross. The latter is permanently involved in bull-dozing large masses of ore--extremely useful bulk material. It is necessarily a social rather than an individual effort. Mathematics can fix its attention on succinct concepts and theorems; computer science can expect, even after equally determined efforts toward simplification, only to build sprawling procedures, which require painstaking and extensive descriptive mapping if they are to be preserved from dusty chaos". If things haven't turned out quite as badly as Schwartz implied, it is because we have succeeded better in abstraction than predicted.

I think that literacy of computing is now so widespread that it makes sense for teachers of mathematics to take advantage of students' prior knowledge in the area. I would speculate that soon many students' first exposure to the notion of _variable_ will be in programming and not in high school algebra. Even if we ignore the computer scientists, it makes sense to make use of this prior experience of the students in mathematics courses. I should say I'm not talking about prior experience in computer science; I'm talking about prior experience in computing. These are very different things.

Discrete mathematics really captures most of the specific subjects that are necessary in computer science. We don't have as urgent a need for calculus. We still encourage students to take calculus because at this point it provides the best introduction to mathematical reasoning that is offered. It is also used in certain advanced computer science courses. But I do hope that we can alter that situation with regard to discrete mathematics; it must be taught as early as possible to computer science students.

Among the themes that should play a role in discrete mathematics courses, induction and recursion are very important. Induction and recursion both capture the notion of reducing a problem and also provide a very nice way of connecting operational or algorithmic thinking with static or relational thinking. If you look at a recursive definition of a function, you can, on the one hand, think of it as a description of the method of computation but, on the other hand, you can think of it as an equation describing a particular function, that is, there is a function that satisfies this equation. Recursion is magical in the sense that it allows you to make this connection.

Weissglass: We've heard from people in other disciplines about their needs. Do you see any possibility of having a unified curriculum that would take care of the needs of computer scientists as well as the needs of engineers, economists and social scientists?

Scherlis: I'm pessimistic about the possibility of a monolithic course. My goal is to find a discrete mathematics course that can serve for computer scientists in the role that calculus presently serves for others. By the way, calculus is by no means irrelevant to computer science; it's just not as primary as discrete mathematics. A lot, perhaps most computer scientists will need some calculus.

Norman: Is linear algebra enough for computer scientists or do they need abstract algebra, too?

Scherlis: The concept of a mathematical system is the most important idea from abstract algebra for the computer science student. They also need certain specific ideas--boolean algebras and semigroups, for example.

DEVELOPING MATHEMATICAL MATURITY

Lynn Arthur Steen
Department of Mathematics
St. Olaf College
Northfield, Minn., 55057

"I can't define it, but I know it when I see it." If Potter Stewart hadn't coined this phrase, mathematicians would have. Trying to define "mathematical maturity" is about as hopeless a task as defining obscenity, or, for that matter, justice or love. Nevertheless, since that is our task, let us begin.

The Oxford English Dictionary describes "maturity" as "fullness or perfection of development." That's actually not at all what mathematicians mean when they use the word. The colloquial use of "mathematical maturity" is reserved for a certain stage in a person's intellectual growth that marks the transition from routine, elementary modes of thought to subtle, complex patterns. This transition ordinarily occurs, if at all, sometime during the college years.

There are several marks of maturity that most mathematicians will instantly recognize. One of the most important is the ability to abstract, to glean the essential structure from a complex situation. The first encounter with algebra is an exercise in abstraction, as is the study of classical Euclidean geometry: each requires of the student a leap of faith in which the security of numbers and objects is abandoned for the formalism of variables and definitions. Analytic geometry, and all the calculus based on it, is another plateau in the student's climb towards powers of abstract reasoning: in analytic geometry, algebra is used to abstract from geometry the essential features for particular applications.

A second mark of maturity is the ability to synthesize, to create new ideas by effective use of old ones. Logical deduction, indirect reasoning, the ability to create and follow an extended argument are key ingredients in synthesis. But so too are the elusive elements of imagination and creativity. There is considerable difference between understanding a proof and creating one, and even more so between proving a conjecture and formulating a conjecture.

These are the two traditional marks of mathematical maturity: abstraction and synthesis. They have very little to do with "fullness of development," the ordinary meaning of "maturity." There is however, a word in the dictionary whose meaning is remarkably similar to what I have just described as "mathematical maturity." It is "intelligence," the ability "to learn from experience, ... to respond successfully to a new situation, ... to use the faculty of reason in solving problems." Abstraction and synthesis are just fancy words for the ability to bring experience and reason to bear on the solution of new problems.

Nature or Nurture

Apart from their different denotations, "maturity" and "intelligence" carry with them considerably different connotations. Maturity is something that one grows into, that is molded by the environment, and that is always subject to improvement. In contrast, intelligence connotes more an innate ability that is only in small ways altered by environment and learning. You either have it, or you don't.

I wonder why we mathematicians choose a word such as "maturity" to define a concept more akin to "intelligence"? Is it perhaps a euphemism coined to avoid confrontation with the reality of inherent ability? The connotation of growth implied by the term "maturity" may be just a masquerade for the blossoming of natural talent.

What's at stake here goes to the heart of the issue at this conference: If the development of mathematical maturity is primarily the result of careful nurturing, then we must be very careful to select a curriculum that provides the proper conditions. On the other hand, if mathematical maturity is primarily an expression of inherent ability, then the curriculum may have only second-order effects on its development.

What little evidence there is supports--but does not prove-- the hypothesis that what we mean by "mathematical maturity" is in reality something more akin to "inherent mathematical intelligence." The most striking evidence is the phenomenon of mathematical prodigies. Individuals such as Gauss, Ramanujan, and Wiener achieved some important aspects of mathematical maturity at very early ages, without any formal instruction in the standard curriculum.

Interestingly, child prodigies are found primarily in those parts of human activity that involve a formal, "unnatural" language: mathematics, music, chess. The early accomplishments of a Mozart or a Ramanujan are qualitatively different from the later accomplishments of a Shakespeare. They are in some human sense "immature," despite their technical virtuosity. They are perhaps too much form without content, expression without meaning. In contrast, the quality of mathematics produced by Hilbert or von Neumann in mid-career is the result of formal virtuosity combined with an incredible insight into a rich variety of mathematical and scientific problems.

This suggests that there are, perhaps, two levels of mathematical maturity: a technical, formal competence that can emerge at almost any age, and a sophisticated, meaning-filled understanding of the relations of mathematics to the world around it that emerges, if at all, only after a long period of growth. The former is really an expression of the ability "to use the faculty of reason," the latter a sign of "fullness of development." It is mostly the former that is on the minds of mathematics teachers when they talk about development of mathematical maturity in undergraduate mathematics.

Let us, then, use the term in this traditional sense, as an expression of growth in the use of mathematical talent--the "faculty of reason." It has been a long-standing objective of education to help an individual develop his or her talents to the greatest possible extent. Insofar as mathematics is concerned, this suggests that the development of mathematical maturity is one of the primary objectives of mathematical education.

Criteria of Maturity

Teachers are eternal optimists, constantly looking for signs of spring in the bleak landscape of winter classrooms. A good question from the back of the room is like a crocus poking its head above the melting snow. It is a sign of emerging maturity. Here are some other signs:

-- The ability to use and interpret mathematical notation.

-- The ability to model, to express real-world problems in mathematical form.

-- The ability to perceive patterns, and to apply principles of symmetry.

-- The ability to estimate, to solve problems by perturbing the data.

-- The ability to generalize, to infer fundamental laws from particular cases.
-- The ability to detect and avoid sloppy reasoning.
-- The ability to see and exploit relationships among various parts of mathematics.
-- The ability to read and understand mathematical writing, whether in mathematics or in scientific contexts.

It is interesting that I naturally phrased all these signs as "abilities." While there might be another suitable word, I couldn't find a better one. "Ability," like "intelligence," denotes innate characteristics more than developmental processes. Looking for evidence of mathematical maturity is like looking for signs of mathematical ability. A few signs, like the crocuses in the snowbank, are just clues of things to come. If the bulbs are in the ground, then in the right season they will all burst into bloom.

Calculus: Intuition of the Infinite

To the extent that mathematical maturity is primarily the expression of latent ability, the debate about whether calculus is the best way to promote its development is moot. Any type of mathematical activity would bring about the desired result. Of course, this is a clear exaggeration of reality, expressing one extreme in the debate over the meaning of mathematical maturity.

The more conventional view is that maturity will emerge only if cultivated in an appropriate, supporting environment. Like a seed that can grow to maturity only in the right kind of soil--dry for some, damp for others--mathematical talent may well require an appropriate curriculum in which to mature.

Realism compels us to recognize that certain types of study reinforce better than others the development of mathematical maturity. In this respect it makes little difference whether maturity emerges from latent ability or is created by sheer force of teacher and student: certain mathematical activities resonate with the characteristics of maturity better than do others.

Let's look first at calculus. There are some compelling reasons for the widespread belief (among both mathematicians and scientists) that calculus develops mathematical maturity better than most other subjects. The fundamental results of calculus represent a virtually unique blend of algebraic, geometric and analytic structure. The topics in calculus reinforce each other in ways that are far more sophisticated than comparable topics in other freshman-sophomore courses. This is due in part to the foundation laid by high school mathematics: calculus is not really a first course, but the climax of a six year long program of study beginning with elementary algebra. Calculus provides the in-depth treatment necessary to complete an intellectual program ranging from freshman algebra to principles of economics, from Euclidean geometry to college physics.

There are two ingredients in calculus that resonate with the characteristics of mathematical maturity: First is the synthesis of fundamental mathematical structures in the concept of a limit, in the completion of the real numbers, and in the fundamental theorem of calculus. Calculus is a deep course, requiring real sophistication for mastery.

Second, calculus represents as does no other part of elementary mathematics what Whitehead called the "infusion of pattern into natural occurences." Calculus is one of the greatest intellectual accomplishments of mankind, the best tool man has ever developed for understanding the physical universe. The intrinsic ability of calculus to model change, whether in astronomy or economics, provides understanding--indeed, maturity--in our perception of the world in which we live.

Calculus in the classroom, however, is not the same thing as calculus in the philosophy book. Where those who think about the nature of mathematics see a powerful intellectual accomplishment, those who teach and learn in a typical freshman course see mostly mindless memorization of patterns. Most calculus courses are just devices for programming people to serve as moderately sophisticated computers. Indeed, the new symbol-manipulating computer programs can solve most problems from typical calculus exams every bit as well as the best students.

Evidence that calculus inhibits rather than promotes maturity can be found in the recurring appearance of books that pretend to help a student bridge the gap between calculus and higher mathematics. Many students--indeed, the vast majority-- finish the calculus sequence unable to read or study mathematics, or to master on their own any part of the traditional core of university mathematics. Sometimes, perhaps more often than we care to admit students with special mathematical aptitude leave calculus with less mathematical maturity than when they entered.

Calculus, really, refers to two things: the ideal, and the real. Any course can be taught in such a way as to repress latent mathematical maturity, or to encourage it. Unfortunately, the massive "calculus industry" all too often smothers the intellectual excitement of calculus, and undermines whatever potential it has for fertilizing mathematical maturity.

Discrete Mathematics: Intuition of the Finite

With respect to the development of mathematical maturity, the case for calculus rests on synthesis: it links mathematics with reality and integrates major areas within mathematics. In contrast, the case for discrete mathematics rests on abstraction. The finite nature of discrete mathematics, its openness to case-by-case investigation, and the arithmetic nature of proof make it much easier for students to investigate patterns and construct proofs.

Abstraction in finite mathematics does not require as big a leap as it does in calculus and analysis. The set of integers is really a lot easier to comprehend than is the set of real numbers. A proof of a beginning theorem in graph theory is much easier than a proof of a corresponding theorem in calculus. Limits are evasive, mysterious, in Bishop Berkeley's words, "ghosts of departed quantities." Vertices and vectors, on the other hand, are specific, graphic, and comprehensible. The gradient from elementary to advanced mathematics is far easier along the road of discrete mathematics than it is along the highway of calculus.

This difference in gradient cuts both ways. While a gradual slope is easier for beginners to climb, it does not achieve as high an elevation during the first two years of study as does the steeper route of calculus. Hence the reluctance of experienced mathematicians to abandon calculus as the preferred route to mathematical maturity: for those who can climb its slopes as college freshman, calculus will do far more than will discrete mathematics. This gap may not be intrinsic to the fields, but it is a realistic assessment of present curriculum practices: calculus builds on the momentum (as Newton put it, on the "shoulders of giants") of algebra and analytic geometry, whereas discrete mathematics in college is starting virtually from scratch. And we must never forget that most students who take one year of college mathematics never take any more.

Nevertheless, discrete mathematics by its very nature has far greater potential for developing mathematical maturity than does traditional calculus. Problems typical of discrete mathematics do not conform as readily to standard paradigms as do the problems of calculus. In discrete mathematics there is a compelling need to engage in careful, logical analysis; geometric intuition, so powerful a tool (or crutch) in calculus, is usually of no use. Methods of solutions are often unique to each problem; templates presented in texts or in lectures rarely do more than provide good hints concerning the solution of related problems. This *ad hoc* structure, which some view as evidence of lack of deep theory, in fact makes discrete mathematics immensely valuable as an opportunity for developing mathematical maturity.

Learning vs. Doing

The development of mathematical maturity depends far more on doing mathematics than on learning mathematics. Doing is very slow, but it does produce maturity. And it is far easier for beginning students to do mathematics in the discrete area than it is in the continuous arena. This is why calculus courses have evolved to be so much learning and so little doing. To be sure, all calculus students do lots of homework; but they rarely do mathematics. Students in computer science, on the other hand, are often actively doing mathematics in the process of designing and implementing complex algorithms. This curriculum contrast is a powerful factor in the present tension between undergraduate programs in computing and mathematics: the beginning courses in computer science can offer students a glimpse of doing new things, whereas the beginning courses in mathematics only offer an endless sequence of exercises, mostly two or three centuries old.

The design and analysis of algorithms, a fundamental theme of both discrete mathematics and computer science, requires the instinct and discipline for case-by-case analysis. (Proofs and solutions in calculus, in contrast, are most elegant when they use general principles that extend easily either to higher dimension or to more abstract spaces.) Because discrete mathematics is amenable to analysis of cases, it provides enormous opportunities to do mathematics, even at an elementary level. It epitomizes many of Polya's heuristics, giving concrete expression to abstract thought.

An excellent example of this can be seen in Seymour Papert's use of turtle geometry to engage primary grade children in doing mathematics. The discrete, algorithmic nature of computers makes them, in Papert's phrase, "mathematics-speaking beings." By removing the impediments to communication, Papert has demonstrated beyond any doubt that children at even very young ages can explore discrete mathematics. Surely no less can be expected of college freshmen.

But to expect college freshmen to do mathematics while studying calculus is almost beyond the intellectual resources of our nation. At least 90% of the students who take calculus depend for their survival far more on memorization than on maturity. They may learn calculus, but they actually do very little mathematics.

Providing Alternatives

The difference between the criteria for success in discrete vs. continuous mathematics bears directly on the issue of mathematical maturity. Because success in calculus depends on applications of general principles, those who master it are well on their way to achieving what mathematicians have traditionally meant by maturity. On the other hand, success at the case-by-case methodology of discrete mathematics is quite a different matter. Not everyone who has achieved some maturity in calculus will be very good at it, nor will everyone who masters discrete mathematics have thereby gained any substantial insight into continuous mathematics. Calculus and discrete mathematics are two fundamentally different areas, different in content, in objectives, and most importantly, in methods of attack.

Donald Knuth recently made a systematic study of the modes of thought of mathematicians vs. computer scientists. He sampled representative classics in each field to determine by empirical means characteristic patterns of thought in each field. In mathematics books he found numerous examples of manipulation of formulas, mathematical modelling, generalization, abstraction, reduction to simpler cases--all instances of our criteria for mathematical maturity. Two features of computer science--information structures and algorithms--were found occasionally in the mathematics literature, but not in especially heavy concentration. And one absolutely vital element of computer science was totally absent from any of the mathematical literature: analysis of the state of a process. Mathematicians simply do not ever set $n = n+1$. They do not make assignments of variables, and thus cannot really capitalize on the dynamic nature of algorithmic processes.

Knuth's study confirms what we already know: individuals differ in their natural instinct for certain types of structure. We all have known students who blossom when immersed in analytic, continuous, physical mathematics, but who have very little interest in or instinct for the algebraic, discrete or algorithmic parts of mathematics. Conversely, our classrooms are now filled with students whose attachment to computing does not entail any special interest in traditional mathematics. (Knuth claims that most of these students do not have any special talent for computing either: according to his figures, only 2% of students have a natural talent for algorithmic thinking.)

This evidence surely suggests that the best strategy for developing mathematical maturity in a large number of students is variety--discrete courses for those who thrive in that environment, and continuous courses for those who prefer this more traditional approach. That's so obvious that it is hardly worth saying. Using it as a conclusion to this paper is tantamount to admitting that the entire argument is just a shaggy dog story.

The clash between analytic and algorithmic intuition, between calculus and discrete mathematics, tangles hopelessly any neat theory of mathematical maturity. Mathematics itself rests on many legs--algebra, calculus, computing, geometry, logic... . As maturity emerges more rapidly in one area than another, the framework of mathematics that emerges in each student's mind shifts its center of gravity precariously and unpredictably. In practice, the growth in use of mathematical talent is very uneven. Mathematical maturity is not a coherent, single entity, but an amorphous mix of diverse characteristics, each supported by special talent and special interests.

I can't define a shaggy dog, but I know one when I see it.

Acknowledgements: I would like to thank Paul Fjelstad, Jennifer Galovich, Paul Humke, Arthur Seebach, and Kay Smith for stimulating discussion that contributed to many of the ideas expressed in this paper.

DISCUSSION

Steen: The normal use of maturity connotes a sense of completeness as when something is full grown. When we speak of mathematical maturity, however, we use this term in the sense of a process of development in which students learn to abstract and to synthesize. Certainly the ability to express problems symbolically is an ingredient in mathematical maturity.

In considering mathematical maturity in relation to calculus and discrete mathematics, support for putting calculus on a separate pedestal is the fact that it is really a major scientific, mathematical and intellectural accomplishment that has deep roots and permeates all of mathematics and science. On the other hand, students don't study calculus in this context; they memorize derivative and integral formulas and plugging in word problems. They don't receive it as a synthesis of algebra and geometry or as a major intellectual accomplishment.

With discrete mathematics students deal with issues - numbers etc. - with which they are more familiar than the fundamental notions of calculus. On the other hand, there is a big gulf between beginning levels and advanced levels of discrete mathematics. Perhaps this distinction is similar to that between calculus and real analysis.

Since most students take only one or two mathematics courses, it's not clear that developing mathematical maturity should be a high priority. Just introducing various types of mathematics might be a higher priority. Therefore, in discussing calculus and discrete mathematics, maybe we should treat mathematical maturity as a second order effect.

Lucas: I believe that the progression from elementary to upper division finite mathematics can be at least as continuous as the calculus, advanced calculus, etc. progression.

Young: Mathematical maturity is really a very odd thing. You see freshmen who have never been close to mathematics who have it. It is something with a strong psychological, not educational component. Calculus is no longer a suitable vehicle at least because the intellectual background which students used to have when they came to calculus is no longer there. Calculus no longer serves to illuminate all the elementary mathematics which came before it.

Wilf: An important ingredient of mathematical maturity is the ability to recognize a proof as opposed to a heuristic argument. But in calculus we rarely see a proof. In discrete mathematics, however, you really can and do prove things.

Bushaw: In a certain sense discrete mathematics now is at a stage similar to calculus in 1660. At least by exposing students to discrete mathematics you give them an opportunity to participate in a contemporary subject which is not slick the way calculus is.

Tucker (Albert): The frontier is very close in discrete mathematics. This is what you mean, I think, when you say that the jump from elementary to advanced discrete mathematics is a very difficult one. It is a great advantage of discrete mathematics that the creative frontier is in sight whereas, with calculus, it is out of sight.

Steen: In good introductory courses in chemistry, physics and, especially, biology students also see the frontier.

Roberts: In discrete mathematics on the frist day of class you can state and then solve problems the students can understand and make some progress on themselves. I don't see any better way of developing mathematical maturity than to get students involved in inventing and doing things themselves.

Maurer: I tend to think of maturity rather narrowly as a ability to abstract from complex situations and to see a conceptual way through problems rather than a messy computational approach. Examples of this kind of insightful thinking are much more accessible to freshmen and sophomores in combinatorics than in analysis. I really wonder though if mathematical maturity is an issue to worry about in considering the first courses in college mathematics. It's been noted that calculus often serves as a screening device for other fields. I believe the reason it screens well (if it does) is that success in calculus requires stamina and systematicity in handling complicated situations. There may be a little bit of danger in discrete mathematics just because it has so few turning-of-the-crank topics.

Steen: The ability of calculus to screen, to sort out is probably why the pre-med curriculum and business schools require it. Will discrete mathematics serve the same purpose?

Norman: Maturity is really a blossoming of the innate. I think discrete mathematics, in which there can be more doing than memorizing, does encourage just such maturity.

Barrett: Mathematical maturity for social scientists may consist only of the ability to read mathematics. We need to distinguish what we mean by mathematical maturity for the mathematician, for the scientist, for the social scientist, etc.

Tucker (Albert): But we must also distinguish between literacy and the ability to do it yourself.

Anderson: Discrete mathematics requires more insight, less manipulative skills than calculus. And so it would serve as a different kind of screen than calculus. We ought to be careful about the implication of this.

Weissglass: There's been an assumption in much of what has been said that it is possible to have a one-year course develop maturity. That seems to be contradictory. Maturity is a gradual process. We need to put attention on what's happening to one-year olds and two-year olds and eight-year olds if we expect them to become mathematically mature, whatever that is. I think it is wrong to think that mathematical ability is totally innate. The potential is there for each, but is is not going to develop unless there is the right environment. The difficulty educators face in developing maturity (or what I like to call functioning ability in mathematics), by teaching one year of discrete mathematics or calculus is that for 12 years students have coerced into rote memorizing without any understanding.

I think the attitude that comes through in Lynn Steen's paper is that some people have mathematical ability and some people don't. This is part of a whole cultural misconception that hinders the development of functioning ability in mathematics. I believe that all our students would develop mathematical ability if we gave them the right stimulation and encouragement starting in first grade or even before. It's not only the curriculum we need to consider but the method that we use to teach.

Steen: There are obvious contradictions in the way we use mathematical maturity and how I use this term in my paper. At the time they enter college, some students are more ready to deal with college mathematics than others. Whether this is because of their prior education or their innate intelligence is probably irrelevant to curriculum strategy in the freshman year.

Kreider: In addition to screening, an advantage of calculus is that students can use it as a measure of their own progress in terms of their persistence, stamina and self-confidence. Students will have to be able to see that finite mathematics can provide similar measures. We must be careful not to lose students to other disciplines because the others look comparatively richer.

Pollak: There is a confusion in the discussion between the maturity of the field and the maturity of the student. In continuous mathematics there are more overall patterns and less need for tricks to handle particular problems than in discrete mathematics. Do we really want to train students to look for the trick?

Maurer: The difference between a trick and an insightful method is a fine one; whether a student will perceive a given solution in one category or the other depends in part on how much motivation the teacher supplies. To teach students only tricks is to miseducate them but to provide them with insights on how to work through problems that require something a little bit different than what they have already learned is important.

Roberts: On tests we can only give trivial problems in either calculus or discrete mathematics. But in problem assignments we can probably use discrete mathematics intelligently to lead students to be inventive and discover things more easily than we can with calculus.

A TWO-YEAR LOWER-DIVISION MATHEMATICS SEQUENCE

D. Bushaw
Washington State University
Pullman, WA 99163

Introduction. There has long been some criticism of the way in which the calculus lords it over the traditional lower-division mathematics curriculum. Growth of "new applications" and the prevalence of digital computing devices have recently been making thoughtful people increasingly uneasy about the relative neglect of discrete mathematics at that level. Their uneasiness can be defended from the standpoint of specific curricular concerns (e.g. the special needs of computer science students) or in general, somewhat philosophical, terms. Much has now been said about the matter, however, and while not all the arguments are equally convincing--some of them seem downright specious--a very good case has been made for displacing most, if not all, of the calculus sequence with topics of a less analytical nature. It is postulated in this essay that moves of this character deserve serious consideration.

There remain formidable questions of detail: what calculus topics might go, what discrete topics might come, to what extent should the new mix be integrated, and so on. Perhaps the best way to attempt to deal with these questions is to propose and consider global solutions: fairly complete course or sequence descriptions that might be judged concretely not only on content but on inner and external articulation.

My main purpose here is to present such a proposal. I have tried to consider the plight of a college or university where it would not be practical to offer numerous sequences of courses in lower-division mathematics for different kinds of students. The ideal is a single two-year sequence which in part or as a whole would satisfactorily meet the needs of almost any qualified mathematics student, whether that student be in pure mathematics, applied mathematics, statistics, computer science, a physical science, engineering, or even the humanities. This is probably an impossible ideal, but the proposal made below might do a rather good job for most of these students and, in those institutions where needs suggest and resources allow embellishments, might still serve as a stem for a richer lower-division offering.

The emphasis in this proposal is on curricular integration. An attempt has been made to arrange topics in such a way that, to some extent at least, discrete and continuous ideas are not segregated but can illuminate and reinforce one another. Besides strengthening the student's experience, this should make possible certain economies of time. Those who discuss the inclusion of more discrete mathematics in the curriculum sometimes speak as if, for every two weeks of discrete mathematics added, two weeks of continuous mathematics must be given up. But just as a cup of sugar and a cup of water do not make two cups of syrup, so two weeks worth of differential equations and two weeks worth of difference equations do not require four weeks of class time. (The analogy, of course, is not perfect; in particular, nothing is implied here about which is the sugar and which is the water.)

If something like the following proposal were adopted, there would immediately arise technical problems: finding suitable textbooks and other instructional materials; further training of some instructors; revision of placement (especially advanced placement) systems; etc. The reality of these problems should be recognized, of course, but they are surely caudal as compared with the primary question: What should lower-division mathematics students be expected to learn?

Assumptions. The following proposal is for approximately sixty weeks of instruction, each week involving from 150 to 250 minutes of class meeting time. The depth to which the topics can be treated and the flexibility of the program will naturally depend in part on where the amount of time actually available falls in this range. The outline is described in terms of five-week units which can be grouped by twos and threes, in various natural arrangements, into reasonably coherent quarter or semester courses.

Students beginning the sequence are expected to have a good secondary school background in mathematics, including: algebra at least through quadratic equations, mathematical induction, and the binomial theorem; geometry through coordinate geometry of lines and the simplest conics; and elementary trigonometry. Further background is desirable, especially if it has not left the student with the impression that he or she has nothing left to learn. Provision should be made for enabling students who have not acquired needed computing skills somewhere to acquire them.

General structure. The five-week units mentioned above can be characterized as follows:

unit	principal topics
1	Infinite sequences, derivatives, continuity.
2	Applications of the derivative; the exponential function.
3	Fundamentals of integral calculus.
4	Multiple integrals; calculus of circular functions.
5	Infinite series and improper integrals.
6	Partial derivatives.
7	Concrete linear algebra.
8	Abstract linear algebra.
9	Elements of mathematical logic; sets and relations.
10	Graphs.
11	Combinatorics.
12	Probability.

As will be clear from the detailed descriptions that follow, a natural dependence diagram for these units might look like:

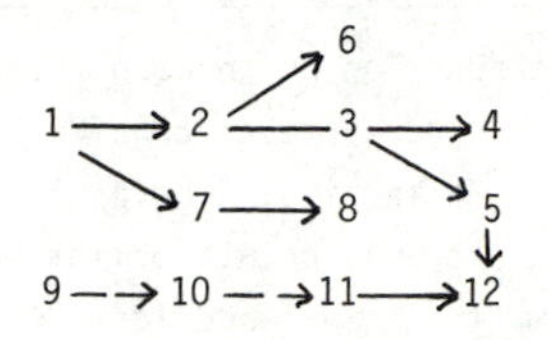

This shows that many arrangements (linear or not) of the twelve units or only some of them are feasible. For instance, the sequence 9-1-7-10-2-3 (one year) might be good for students in social sciences or business. It is to be assumed that a student concentrating in mathematics, computer science, or physics, for example, would take all twelve units and significantly more.

The twelve units are described in detail below. A few topics of marginal importance are given in parentheses. Topics that seem especially suitable for computer treatment will be marked with an asterisk. Beyond this, little will be said about computer or calculator implementation, mainly because equipment and software vary so much from school to school and evolve so rapidly. There will be many opportunities throughout the units to point to examples of algorithms. Enough of these opportunities should be taken to acquaint students with "the algorithmic point of view" in general, and somewhere the elements of the theory of algorithms should be introduced. In the same spirit, some time should be set aside for such matters as: special problems arising when arithmetic is done on calculators; a review of induction; and the nature of mathematical modeling.

The Units.

Unit 1: Infinite sequences, derivatives, continuity.

Concept of an infinite sequence; examples defined explicitly and iteratively*; algebraic operations on them. Convergence and limits*. Definition of differencing; application to polynomials*; higher differences; introduction to summation. Definition of derivative by sequences* (probably preceded by review of functional notation). Interpretation as slope, velocity, temperature gradient, marginal cost, etc. Algorithms for differentiation of polynomials, products, and quotients*; higher derivatives. Definition of continuity by sequences*; graphical interpretation; major types of discontinuities and how they arise in simply defined functions. Limits of functions of a real variable; preservation under elementary operations. [Comments. The sequential approach to differential calculus has been tried before and did not catch on, probably for reasons of little significance to beginners. It is easy, lends itself to computational illustration, and helps to set a "discrete" tone for the whole program. Students are likely to find differentiation by symbolic manipulation on a computer exciting, but this possibility naturally does not make some facility in differentiation "by hand" obsolete.]

Unit 2: Applications of the derivative; the exponential function.

Maxima and minima of a differentiable function on an open interval; critical points; sufficient conditions by first or second derivative; consideration of points of non-differentiability, ends of interval of definition. Curve plotting by points*, use of information about discontinuities, extrema, points of inflection (define), simplest kinds of symmetry. The theorem of the mean and its use in estimating differences*; differentials as approximations to differences. Inverse functions and their derivatives. Newton's method; basic recurrence relation motivated graphically; examples*; use to find extrema*. Partial derivatives and their interpretation. Sequential definition of exp; its elementary properties and derivative. Solutions of $y' = ky$; this equation as model of growth or decay. The natural logarithm, its properties and derivative. (Hyperbolic functions.) [Comments. This is mostly important, bread-and-butter material, appealing to almost all prospective calculus users; and the appeal will be enhanced by offering a balanced mixture of formal algorithms, calculation, and graphical techniques, including judicious use of CRTs or plotters. Here and elsewhere, theoretical subtleties and possible pathologies should not be overemphasized.]

Unit 3: Fundamentals of integral calculus.

Approximation of plane areas by summing areas of thin rectangles; the trapezoidal rule*. Definition of the definite integral (analogy with sequential definition of limit in unit 1); simple examples of definite integrals calculated directly*. Theorem on existence for continuous functions on compact intervals (no proof). Two basic forms of fundamental theorem; antiderivatives; their use to evaluate definite integrals. Techniques of integration: tables, substitution, parts. (Integration of rational functions.) Simpson's rule. Areas and volumes by cross sections. Average of a continuous function over an interval*. Applications such as work, cumulative income*. [Comments. As compared with the conventional course, this unit is cut drastically. In the present intellectual environment, it is indefensible to expect all calculus students to become virtuosi in analytic techniques of integration or to master such archaic and rather artificial applications as the force on a dam. One reason for the traditional emphasis on techniques of integration has been to prepare students for the conventional first course in differential equations, most of the problems of which are now probably best approached by numerical or qualitative methods. The semester's work in units 1-3 could constitute a terminal "short calculus" of the type now often required of certain students. Indeed, it at least mentions the fundamental ideas of the discrete as well as of the continuous calculus, and sheds some light on relations between them.]

Unit 4: Multiple integrals; calculus of circular functions.

Rectangular coordinates in 3-space; sketching simple loci there*. Multiple integrals, repeated integrals, and the relations between them. Computation* of double and triple integrals over fairly well-behaved domains. Review of circular functions: definitions, basic identities, values at selected angles. Derivatives and integrals of sine and cosine. Remark on treatment of other circular functions in terms of these. Examples of trigonometric substitutions. (Inverse trigonometric functions and their derivatives and integrals.) Polar coordinates; integrals in polar coordinates. Brief discussion of cylindrical and spherical coordinates. [Comments. Here again considerable streamlining of the usual fare is intended; what remains will be essential for some students, totally expendable for others. The new information in this unit may be used to extend the range of units 2 and 3; for example, students should now be asked to do exercises involving extrema of functions in which circular functions appear.]

Unit 5: Infinite series and improper integrals.

Definitions of infinite series, partial sums, convergence. Examples and counterexamples*. Alternating series, comparison and ratio tests. Absolute convergence, permutation of terms. Power series and their convergence*. Taylor's theorem with remainder. Manipulation of power series; their uses in approximation* and with generating functions. Termwise differentiation and integration. Brief discussion of improper integrals: definitions, convergence, examples and counterexamples*. [Comments. This is not far from the standard coverage, and perhaps the most notable thing about the unit is the relatively large number of asterisks. Indeed, this area is fertile ground for experimental calculation, which can now be performed with speed and accuracy almost unthinkable not long ago. While no substitute for the basic definitions and theorems, this possibility is a great aid to the strengthening of intuition about "infinite" processes.]

Unit 6: Partial derivatives.

Review of definition of partial derivative. The chain rule. Examples and applications (e.g. "related rates" and implicit differentiation). The total differential as an approximation to an increment and as another way of expressing the chain rule. Taylor series of functions of several variables. Critical points of functions of several variables*. Directional derivatives and gradients. Level sets, tangent hyperplanes. The idea of a vector field, with illustrations. Line integrals and Green's Theorem. [Comments. From the fact that this material is to be covered in five weeks it follows that some of it, namely aspects of vector analysis often presented at some length at this level, must be treated briefly if at all. In fact, the ideas of vector analysis are important to far from all the students, and for those for whom it is important it is usually covered *ab initio* and more fully in later courses.]

Unit 7: Concrete linear algebra.

Systems of linear equations; solution by Gaussian elimination*. Matrices, vectors, and operations on them. Linear independence. Definition and calculation* of matrix inverses, determinants defined by cofactor expansions*; Cramer's rule*. Eigenvalues and eigenvectors*. At least one of the applications: linear programming, linear ordinary differential equations, linear recurrence relations. [Comments. This is the earthier half of the usual sophomore linear algebra course, is the first of the almost strictly "discrete" units, and depends on earlier units only through the application to differential equations, which may be avoided. Unit 7 thus cut free could be offered much earlier in the sequence. However, students seem to have unexpected difficulty with this material and that in unit 8 (and these two units should probably be taken successively by those who take them both), so it is best not to rush students into them.]

Unit 8: Abstract linear algebra.

Euclidean n-space (vectors, operations, dot product, norm). Linear spaces over R; subspaces; basis, dimension, rank. Inner products, orthonormal bases, Gram-Schmidt orthonormalization*. Coordinates, change of bases*; linear transformations, kernel, range. Matrices of linear transformations; diagonalization. [Comments. This is most of the rest of the second-year linear algebra that has become standard in recent years. There could easily be too much material here for five weeks if too much time is spent on any one topic.]

Unit 9. Elements of mathematical logic; sets and relations.

Propositional connectives defined by truth tables; propositional calculus; Boolean functions and a glimpse of the problem of simplifying them. Existential and universal quantifiers and their properties. Sets, inclusion, operations. Some basic identities. The set builder notation. Cartesian products. Relations; composites, inverses; images and inverse images. Equivalence relations and partitions. Partial order, linear order. Some elements of cardinal number theory. [Comments. Many students will have learned some of these ideas before coming to college. The new ideas should be abundantly illustrated from everyday life, from disciplines in which the students are interested, and from mathematics courses they have taken. Material of this kind has been neglected in American college mathematics curricula, but because of its pervasiveness in mathematics, its special pertinence to computer science and other young fields, and its fundamental character, surely deserves an important place in the curriculum. This unit could be a good place for a review, with enrichment, of induction.]

Unit 10: Graphs.

Graphs as relations and as models; chains, paths, and connectedness; isomorphism. Trees: basic properties; use in searching; breadth-first and depth-first search*;

spanning trees. Euler circuit theorem and extensions; hamiltonian graphs; applications of eulerian and hamiltonian circuits*. [Comments. The outlines for this unit and the next are adapted from the CUPM Recommendations for a General Mathematical Sciences Program (1981), pp. 27-8. Experience with teaching graph theory at this level is probably rare, so more than usual tentativeness in the choice of topics is in order. The general subject is an excellent one to include, however, not only because of its intrinsic importance in the mathematical sciences but because of the breadth and diversity of its applications, which makes it an unusually good source of brief and easily accessible examples of mathematical modeling.]

Unit 11: Combinatorics.

Counting based on tree diagrams; counting by decomposition into subcases. Permutations and combinations*, binomial coefficients, Pascal's triangle; multinomial coefficients; selected identities. The inclusion/exclusion principle and modeling with it; derangements; graph coloring. Generating functions. [Comments. See the "Comments" on unit 10 above. Here again there are many opportunities for bringing in illustrative applications, not only after the applicable mathematical machinery has been developed but before, to arouse students' interest. Almost every item in this description could have been marked with the asterisk indicating opportunities for computer or calculator use.]

Unit 12: Probability.

Axioms and basic rules, independence. Effective calculation of discrete probabilities*. Expectation. Univariate density and probability functions; moments; law of large numbers. Binomial, Poisson, exponential, and normal distributions. Central limit theorem (no proof). Conditional probability and Bayes' theorem. [Comments. This description also owes much to the CUPM Recommendations, p. 98. Here again, an asterisk could have been put on almost every word. For example, as the Recommendations suggest, computer simulation could be used to illustrate the law of large numbers and the central limit theorem. Because of its position at the end of the list, this unit could be changed considerably without necessitating changes in other units. In some programs, for example, it might be considered desirable to suppress the discrete aspect almost entirely; in others, it might be considered necessary or at least wise to make the unit accessible to students who have not studied calculus, and thus to concentrate on discrete probability. In either case an introduction to basic statistical ideas might be considered more important than some of the topics listed. Large schools could probably offer courses including probability units of several types, thereby catering more precisely to a wider range of students.]

Conclusion. Much more could have been said in the preceding "Comments" about: reasons for listing the twelve units in the order given; reasons for including some topics and excluding others; degrees of usefulness of individual units to students drawn from different constituencies; details of coverage and manner of presentation;

and so on. For experienced mathematics instructors, many of these things will be clear. In some respects, however, each of us is inexperienced, and brevity may be a sign of ignorance or (to put the matter more positively) an invitation to experiment.

Indeed, the tenative and experimental nature of any such proposal can hardly be overemphasized. The usual lower-division mathematics curriculum, in its many variations, is the product of a long evolution and has been tested by many students and teachers, in many forms, under many different sets of circumstances--and still is vulnerable to rather far-reaching criticism. Nothing closely resembling the sequence proposed in the present essay has ever, to my knowledge, been taught anywhere, and it therefore almost certainly contains gaps, bugs, and excesses, some of which might not come to light until it has been taught, perhaps several times. The same might surely be expected of any such proposal made _in vitro_, if not _in vacuo_. What is needed is a long process of serious but good-natured development, comparison, criticism, testing, analysis, and synthesis of new ideas. This essay is intended as no more than a modest contribution to this difficult and endless process.

DISCUSSION

Bushaw: We should not want or hope to produce a hard and fast outline for lower division mathematics courses. If we produce a plausible set of recommendations and if the people behind them have some credibility, then textbook writers will be influenced and, in turn, instructors of mathematics will adopt these textbooks.

My aim in developing an integrated two-year curriculum was to preserve the essentials of calculus while inserting a fair amount of discrete mathematics.

Wilf: What do you think would be the difference in students' perceptions between the course you propose and two separately packaged courses, one in calculus, one in discrete mathematics?

Bushaw: It seems to me that integration of the two themes gives you more frequent and easier opportunities to allow them to act on one another. The differential equations - difference equations parallel is just one example. Discrete and continuous mathematics are, if not fraternal twins, at least siblings.

Barrett: Have you ever tried teaching pieces of this in five week packages?

Bushaw: I do think in these terms in laying out my courses and I tend to test on five week packages.

Alo: Does your proposal imply that you should have separate tracks for different areas like the social sciences or business students or even engineers?

Bushaw: Yes, although some groups like mathematicians, computer scientists, engineers and physical scientists should probably know all of this.

Norman: Wouldn't different orderings and different subsets for distinct groups cause a lot of backtracking for students who change their mood about their major?

Bushaw: It's a possibility but I'm not sure that its a danger. Students who are convinced to change their major usually expect to make some sacrifice.

Steen: The apparent flexibility in your model just won't be available at smaller institutions which must make a commitment to a particular sequence.

Bushaw: The flexibility would be available to the department not the student. The units in my paper would not necessarily be called to the attention of the students at all. They are just elements from which the faculty might assemble courses.

Tucker (Alan): A strength of your proposal but also a weakness is that, in some sense, it is just adaptation of what isn't so unusual nowadays in a lot of schools. In the CUPM Mathematical Sciences recommendations there was a year of single variable calculus and then we talked about intermediate courses such as linear algebra, discrete mathematics, and multivariate calculus. What you have done really is squeeze six semesters of the current curriculum into six quarters. This is the strength but the weakness is that you haven't attacked the fundamental role of calculus at the beginning.

Bushaw: But what I have numbered 1 to 12 need not be given in that order.

Lucas: Twenty years ago at the University of Michigan we were teaching something very close to this except for the abstract algebra and graph theory. The people who wanted additional differential equations just took it later.

Pollak: Another thing missing besides differential equations is statistics. But overall I think this is a wonderful thing to work from. Your directed graph probably would be better as an undirected graph which would have some loops which don't show up here but which we'll find when we go into it in more detail.

Norman: Mathematical induction is missing, too.

Bushaw: Early in the paper I assume this has been done in the high schools but really you can't make this assumption.

Greber: I think your structure is a completely workable system. Engineers need and would take more mathematics, say six semesters, which would include differential equations.

HOW TO CURE THE PLAGUE OF CALCULUS
(OR REVISIONS IN THE INTRODUCTORY MATHEMATICS CURRICULUM)

a discussion paper

Fred S. Roberts
Department of Mathematics
Rutgers University
New Brunswick, New Jersey 08903

1. The Fundamental Ideas of the Paper

The story goes that calculus was invented because of the bubonic plague. Indeed, the Great Plague of 1664-65 closed Cambridge University, and led Isaac Newton to retire to meditate. As a result, he invented the calculus ([2]). Now many of our students would not be too surprised to learn that calculus was born of the plague; indeed, they rather think of it as a plague in its own right. The fundamental premise of this paper is that the plague of calculus needs curing. Indeed, I shall argue that today's introductory mathematics curriculum, with its almost exclusive emphasis on calculus, is short-changing and misleading students. Then, I shall try to describe how we can do better.

The major damage in college mathematics is done in the first two years of the curriculum. Let us see why that is true. The first two years of the curriculum in mathematics currently consist of four semesters of calculus with some linear algebra. Now what roles should this introductory curriculum play? First, it should introduce students to what modern mathematics is all about, whether or not they will go on with mathematics. Second, it should provide students with the specific tools they need to take more mathematics courses. Third, it should provide (non-math) students with the tools they need to take non-mathematics courses and with examples to appreciate the role of mathematics in their discipline and to understand what additional math courses might be relevant for them. If the curriculum does not perform these roles well, then it misleads students, sends potential math students off into other disciplines, fails to attract other students to more math courses which are relevant to them, and in general leaves students unprepared and dissatisfied. All this is happening.

Let us consider the first role. The freshman-sophomore mathematics curriculum should serve as a survey of modern mathematics, indicating its variety, its power, its usefulness, and its beauty. This survey can either be an end in itself, or it can be thought of as an entry point for further study, delineating the possible directions, and helping a student to see what the major areas of modern mathematics are all about. Finally, this introductory curriculum can play a combined role, for the student who starts out wanting just a survey, but might end up getting excited by one aspect of modern mathematics or its applications.

It is my feeling that as a survey of modern mathematics, the present introductory curriculum falls very short. It omits even the slightest mention of some of the most rapidly growing areas of mathematics. I am thinking specifically of discrete mathematics, probability, and linear programming, and, to a lesser degree, of computational and numerical methods. Hence, this introductory curriculum, with its emphasis on continuous mathematics, gives a biased view of what mathematics is all about, and omits some of the topics which might be most interesting and useful to the students.

As for serving as an entry to the math major, or to further courses in mathematics, the present introductory curriculum falls short again. Many important junior and senior math courses have very little to do with the first two years' curriculum. These courses include linear optimization, operations research, modern algebra, combinatorics, probability, statistics, graph theory, mathematical modeling, and of course computer science courses. The first two years' curriculum should introduce a student to these other areas and show him how broad, useful, and fascinating modern mathematics is. Instead, this curriculum presents a narrow, one-sided view. As a result, many students who might otherwise continue with mathematics lose interest.

For the same reason, the introductory curriculum is not fulfilling role number two very well. It does not provide the student with the tools to do well in those more advanced courses taken by an increasing majority of students. Moreover, without any prior exposure to these other areas, the first exposure is forced to be much more elementary than need be. For instance, a student with a good introductory linear algebra course in the first two years, could take a more serious linear algebra course in the junior year, or a serious linear optimization course.

As for the third role, namely providing appropriate tools and examples for non-math majors, the freshman-sophomore curriculum is again falling down for many students. This is especially true for students in computer science, and for many students in biology, economics, and other subjects. Although these students need topics from the calculus, they often don't need them as early, or it is better for them to be exposed to other topics as well. A computer science student who does not see discrete mathematics topics early on is being deprived of a crucial tool and prerequisite for later courses. An economics student who does not see linear algebra and linear programming early on, and some discrete math, is missing something very critical to his field, and is perhaps not getting the full picture of why mathematics is so important for his field. He may not see the need to study mathematics further. A biology student who does not see statistics, numerical methods, or probabilistic ideas early on, is missing some of the most important topics for modern-day biology.

In what follows, I try to make suggestions about how to modify the current freshman-sophomore curriculum in mathematics. It should be noted that the required modifications will have to differ somewhat depending on the interests of the students.

For instance, students interested in computer science will have different needs from students interested in physics. It is the necessity for the first two years' curriculum to serve as a tool for more advanced courses in a variety of fields that places the greatest obstacles in the way of potential revisions. In particular, for students interested in the physical or engineering sciences, these obstacles loom large. However, even physics or engineering students need some early exposure to "modern" topics such as probability, statistics, linear algebra, and some ideas of discrete math. Indeed, many engineers (e.g. industrial, electrical) need to be exposed to many of the same topics as computer scientists. And even physicists and mechanical engineers are being introduced to ideas of probability in their own courses, and need enough discrete math to become conversant with the computer.

Ideally, the first two years' curriculum would be different depending on the orientation of the student: is he interested in physics or engineering, in biology, in economics, in computer science, or in abstract mathematics? Or is he simply trying to get a liberal arts view of what mathematics is all about? Unfortunately, many institutions do not have the resources to create many different introductory mathematics sequences. Also, if the introductory courses are too specialized, and not interchangeable, then a first year student will be forced to make a decision at an unfortunately early point in his college career.

In what follows, I will describe a first two years' curriculum which I feel fulfills well the three roles I have described. All students will start with the same first course. The curriculum will have several "tracks" through it, and a relatively early, but not irrevocable, decision point for students in the middle of the freshman year. One of the tracks, that recommended for physics and engineering students, will not be substantially different from the current one. All other tracks will depart to some extent. The departures are, in my view, very modest compared to what many people are suggesting. Nevertheless, it is my feeling that this curriculum, with its greater scope and its great flexibility, should go far toward curing the plague that is the current introductory calculus sequence.

2. General Description of the Introductory Curriculum

The following eight courses, to be described in detail below, make up the courses from which a curriculum for the first two years will be chosen: Calculus 1,2,3, Computer Science, Discrete Mathematics, Linear Algebra, Differential Equations, and Probability/Statistics. The course in Computer Science will hopefully not be counted as a Math course. Calculus 1 will include discrete topics (see below). All students would take Calculus 1 and 2, Linear Algebra, and Computer Science in their first two years. Moreover, all students would be encouraged to take Discrete Math in the first two years. Most students would take either Calculus 3 or Probability/ Statistics to round out their first two years. Students in Physics and Engineering might defer Probability/Statistics and take Differential Equations instead. Although non-calculus courses will replace some calculus courses in the first two years, it is intended that these new courses be taught at an intellectual level comparable to the introductory calculus sequence. By the end of their college career, almost all students will have taken all eight basic courses. In addition, many students would want to replace the single course in Computer Science by a two course sequence, and others would want to replace the course in Probability/Statistics by a two course sequence. Five representative programs are shown in Table 1.

Table 1: Representative Programs

Program A

	year 1	year 2
semester 1	Calculus 1 Computer Science (can be deferred to second semester)	Calculus 3 Linear Algebra
semester 2	Calculus 2 Optional: Discrete Math	Differential Equations Optional: Discrete Math or Probability/Statistics

Program B

	year 1	year 2
semester 1	Calculus 1 Computer Science (can be deferred to second semester)	Linear Algebra Discrete Math
semester 2	Calculus 2	Calculus 3 and/or Probability/Statistics

Program C

	year 1	year 2
semester 1	Calculus 1 Computer Science (can be deferred to second semester)	Calculus 2 Linear Algebra
semester 2	Discrete Math	Calculus 3 and/or Probability/Statistics

Program D

	year 1	year 2
semester 1	Calculus 1 Computer Science (can be deferred to second semester)	Linear Algebra and/or Probability/Statistics
semester 2	Calculus 2 Discrete Math	Calculus 3 and/or Probability/Statistics

Program E

	year 1	year 2
semester 1	Calculus 1 Computer Science (can be deferred to second semester)	Linear Algebra
semester 2	Discrete Math	Calculus 2

I envision that all students will start with the same first course, which I call Calculus 1, but which in reality contains more than just calculus. Some people will want to make this first course much less calculus than I am proposing. This is certainly possible, and maybe even desirable, for many students. The advantage of the proposal I am making is that the same first course can be used by essentially all students. In the second semester of the first year, students will branch, to either Calculus 2 or a course in Discrete Math. Students interested in the physical or engineering sciences would definitely choose Calculus 2, though they would be encouraged to try to take the Discrete Math course also. Most other students would have a choice. (More on how to make the choice below.) In addition, first year students will be encouraged to take a course in Computer Science in either the first or second semester, and perhaps a one year sequence in Computer Science.

The second year student interested in physics or engineering would take a fairly traditional program, completing Calculus 3, Linear Algebra, and Differential Equations in the sophomore year. Such a student would be encouraged to complete Discrete Math in the first two years, but could defer this and Probability/Statistics to the junior year. Such a student's program is called Program A in Table 1.

Almost all other students would definitely sample Discrete Math in the first two years. They would all take Linear Algebra in the Sophomore year. And they would almost all complete either Calculus 3 or Probability/Statistics during that year as well.

How would a student decide which type of program to take? The first big choice will come in the second semester of the first year, when a student will have to choose to continue with calculus or take a course in Discrete Math. Now I recognize that this is very early to make a student make an important choice. However, the curriculum I describe is flexible enough so that the choice will not be irrevocable. The choice will be made easier by modifying the traditional Calculus 1 course to include some discrete topics. A student will then be able to decide whether to continue with calculus or sample Discrete Math, on the basis of how much he enjoyed the calculus topics as opposed to the discrete topics in Calculus 1, on the basis of his performance in Calculus 1, and on the basis of his potential direction of specialization.

3. Description of the Courses

I will close by describing the eight basic courses.

CALCULUS 1 (with Discrete Math)

This course will aim to introduce the students to many of the traditional topics of the first calculus course, while at the same time introducing them to enough discrete topics to open up their horizons to other types of mathematics, and to help them make their choice of a second semester course. The course should emphasize applications from a variety of disciplines, begin introducing the student to the idea of a mathematical model, and start to emphasize the notion of algorithm.

The course will begin with an application. The application should be interesting, easy to describe, and lead to a model whose solution requires an algorithm. Ideally, the model should lead to open problems which will interest the students. All of this can be done in one lecture, with some simple discrete problems such as the one-way street problem. (See [4].) Next, the course will continue with elementary set theory (set, subset, union, intersection, Cartesian product, etc.). It will introduce binary relations and graphs and digraphs defined from binary relations, emphasizing applied problems as the source of relations and graphs, and, ideally, picking up on the problem introduced in the first lecture. The course will review Cartesian coordinates and elementary analytic geometry. It will introduce the notion of function as a special type of relation and also as related to a computer which computes an output for every input. The course will proceed through the study of many special functions of significance for calculus: linear functions and their graphs; polynomials, and in particular graphs of the polynomials $ax^2 + bx + c$; periodic functions, in particular trigonometric functions.

The course will next introduce the notion of sequence. Sequences will be defined not just by equations for the n-th term, but also by means of recurrence relations. Notions of induction and recursion will be discussed. Solution of recurrence relations by iteration will be described, and the relevance to computation will be stressed.

The notion of limit will be introduced in the context of sequences. As a good application, the notion of limit will also be used to define other functions of significance to calculus, in particular power functions such as $y = 2^x$, and the relevance of such functions to exponential growth (of population, capital, etc.) will be discussed.

Limits of sequences will lead to limits of functions, and from there to continuity and derivatives.

The next part of the course will be quite standard calculus: derivative of x^n, product and quotient rules, chain rule, derivative of sin and cos, inverse functions, and implicit differentiation. However, relevance of the derivative concept in other disciplines should be stressed (e.g., through a discussion of velocity, marginal cost, etc.).

Next, there will be the traditional discussion of curve sketching, critical points, and first and second derivative tests, perhaps with less emphasis than in the past. Emphasis will be placed on related rates and maximum-minimum problems, with applications chosen from a variety of disciplines. This is a good place to reinforce the graph theory, for example by computing the maximum number of edges in a disconnected graph of n vertices, using calculus. This course should seek to integrate discrete and continuous topics as much as possible.

The course will then introduce anti-derivatives and indefinite integrals. The differential equation $y' = ky$ will be discussed, with applications to growth and decay.

The course will probably end by pointing out how area is computed using an indefinite integral. It will probably not have time for a detailed treatment of the definite integral. Perhaps the connection between area and the indefinite integral could be demonstrated by estimating some areas by computer or calculator and showing how the result is close to that obtained using the indefinite integral.

The course will probably not have time for a careful discussion of Riemann sums, the mean value theorem, or the fundamental theorem of the calculus. Applications dealing with the notion of average value (average concentration, average price, etc.), with computation of volumes by cross-sections and disks and by cylinders, with arc length, and with density, weight and center of gravity, will have to be omitted to make room for the topics which have been added. These topics can be mentioned, especially in a course aimed at physicists and engineers, but their detailed treatment awaits Calculus 2.

One of the biggest dilemmas I face in describing this course is my desire to include in it a discussion of exponential and logarithmic functions. The course will introduce power functions $y = 2^x$, as noted earlier, and could discuss logarithms there too. However, there seems not to be time for more. Introduction of the natural log, e^x, and the derivatives of these functions, will be studied in Calculus 2.

The amount of time spent on discrete topics in this course can be varied. It should be at least two to three weeks. However, a more substantial discrete component could be as long as, say, five to six weeks, and still allow adequate time to cover the calculus topics.

CALCULUS 2

When it comes to this course, I subscribe pretty much to the report of the Calculus Subpanel of the Tucker Panel (the CUPM Panel on a General Mathematical Sciences Program - see [3]), except that that report does not envision deferring introduction of the definite integral to Calculus 2. This course studies the following: the definite integral and its applications to area; applications of integration (average value, volume, arc length, density, weight, center of gravity, etc.); the mean value theorem and the fundamental theorem of the calculus; logarithms and exponential functions, their derivatives and integrals, and their applications; techniques of integration; elementary differential equations (separable equations, first order linear equations, second order linear equations with constant coefficients); further topics in sequences; series. This course should differ from a standard Calculus 2 course by introducing numerical techniques and computer applications and by putting emphasis on such concepts as error estimation, rate of convergence, and round-off error. The course should also emphasize modeling problems and applications from a variety of fields. Some UMAP-type modules should be useful here. The course can considerably shorten the traditional discussion of integration, given the increasing availability of computational aids. If this course is routinely deferred to the second year for many students, it could be made a higher-level course than the usual first year calculus course, at least in the sections aimed at math majors. In this case, the course might begin with a more careful review of limits, continuity, and derivatives.

CALCULUS 3

This is the multivariate calculus course, which treats such topics as the analytic geometry of functions of several variables, the notion of partial derivative, and the concept of multiple integral. For some students, in particular students in economics, functions of several variables are more important than many of the topics in Calculus 2. If an institution has the flexibility, it should consider designing a special Calculus 2-Calculus 3 sequence for students in economics, which places topics such as partial derivatives in Calculus 2, and defers topics such as sequences and series to Calculus 3. Thus, students taking only two calculus courses will be exposed to the topics of importance to them.

The linear algebra required for this course should be minimal. Much of the linear algebra often contained in the Calculus 3 course should be put in a Linear

Algebra course. The Calculus 3 course will need only some simple concepts of vectors in 2 and 3 dimensions, the notions of dot product and cross product, the definition of a matrix, and the computational definition of a determinant using minors. I do not think an abstract treatment of multivariate calculus, using linear transformations, is appropriate for most students.

DIFFERENTIAL EQUATIONS

I subscribe to the recommendations of the Calculus Subpanel of the Tucker Panel [3] here. Many topics in differential equations should be covered in earlier courses. For instance, we have already pointed out that the differential equation $y' = ky$ will be discussed in Calculus 1, and that Calculus 2 will place considerable emphasis on differential equations. Moreover, the Linear Algebra course to be described below should include as an application the solution of linear constant coefficient systems of differential equations using eigenvalue methods.

Thus, the differential equations course can be deferred by most students to the junior or senior year. It is a highly recommended course for most math majors.

DISCRETE MATHEMATICS

This course is very similar to the Discrete Methods course described by the Tucker panel [3]. Discrete Mathematics is perhaps the fastest growing area of modern mathematics. It is closely related to the growth of computing, and also to applications in diverse fields. A Discrete Math course should introduce the student to this subject and to its many applications. The course should introduce graph theory, combinatorics, and discrete probability. It will emphasize the notion of algorithm, and use this notion to interrelate fundamental ideas. The course, as I envision it, will start with graphs as models of real-world phenomena, giving many examples. It will study connectedness properties of graphs and digraphs. It will introduce the notion of tree and spanning tree, and will include basic tree-based algorithms such as depth first search and breadth first search. The course will present basic results and applications of graph coloring and of eulerian and hamiltonian circuits, stressing applications to scheduling, coding, genetics, etc.

Next, the course will turn to combinatorics. It will introduce basic counting principles, such as the sum and product rule, and it will introduce permutations, combinations, binomial coefficients, and multinomial coefficients. Applications should be emphasized. Computation of run-time or storage-space for computer algorithms should be a major motivation for the interest in counting. To tie in with the first part of the course, counting techniques will be applied to graphs, for instance in counting and enumerating graphs and in calculating simple chromatic polynomials. At this point, elementary discrete probability notions will be introduced, and counting techniques will be applied. However, the course will not go as far as independent events and conditional probability.

Next, the course will pick up on the discussion of sequences in Calculus 1, emphasizing recurrence relations. It will study the general solution of homogeneous linear recurrence relations and will study such topics as the Fibonacci numbers and their applications.

The course will conclude by studying particular counting techniques, such as inclusion/exclusion. It will apply inclusion/exclusion to the study of derangements, chromatic polynomials, and so on. The course might have time to introduce some optional topics, such as simple experimental designs (Latin squares, orthogonal Latin squares, balanced incomplete block designs), error-correcting codes, or flows in networks.

LINEAR ALGEBRA

The material of linear algebra is fundamental to much of modern mathematics, in particular applied mathematics. This course will introduce the basic ideas of linear algebra, i.e., vectors, matrices, determinants, vector spaces. It will emphasize the applications of these topics to such fields as operations research, statistics, economics, and differential equations. Although the traditional emphasis in the mathematics curriculum has been on linear equations, this course will treat linear inequalities as equally fundamental. One of the goals will be to introduce the student to linear programming, which every mathematics student should be exposed to early in his college career.

The course will begin with vectors, matrices, and matrix operations, using economic and other examples to motivate the concepts and definitions. Then the course will consider solutions of systems of linear equations, again giving many applications. It will introduce Gaussian elimination, and then apply it to computation of the inverse of a matrix. At this point, applications of Gaussian elimination or matrix inversion should be introduced. These include least squares fit; finite Markov chains; or linear economic models (the Leontieff model).

Next, the course will introduce linear inequalities and systems of linear inequalities, and apply them to linear programming, through the geometric solution.

From here, the course will define determinant and study properties of determinants, Cramer's rule, and computation of determinants.

Vector spaces will be introduced, first concretely by studying R^n. Linear independence, basis, and dimension will be studied at a relatively elementary and concrete level, with an emphasis on computation. The generalization to abstract vector spaces should be briefly mentioned, with simple examples given. However, there will probably not be time to discuss linear transformations.

Next, the course will introduce eigenvalues and eigenvectors, and will emphasize computation of eigenvalues and eigenvectors and their many applications. The problem of finding a diagonal matrix similar to a given matrix will be discussed (with the observation that an $n \times n$ matrix is diagonalizable if and only if it has n linearly independent eigenvectors).

Finally, the course will study systems of differential equations $X'(t) = AX(t)$, where A is diagonalizable, and will use the results on diagonalization to find solutions. Applications of such systems of differential equations will be included.

PROBABILITY/STATISTICS

The Subpanel on Statistics of the Tucker Panel [3] recommends that, ideally, a student should have a one-year sequence in Probability/Statistics. In cases where this is difficult, a one-semester combined course is described. This combined course covers obtaining data (random sampling, experimental design), organizing and describing data (tables and graphs, univariate and bivariate descriptive statistics), probability (axioms, independence, random variables through the law of large numbers, and standard distributions such as the binomial, Poisson, exponential, and normal, through the central limit theorem); and statistical inference (random samples, tests of significance, point estimation methods, confidence intervals, and simple linear regression). This course makes sense if very little is proven, for otherwise there is not enough time to cover all these topics. In an elementary course, omitting most proofs is appropriate. It also must be arranged (as is assumed and as I recommend) that the discrete math course cover combinatorial enumeration problems and discrete probability. Moreover, a later course in modeling or operations research should cover stochastic processes, at least finite Markov chains. (This topic might also be covered in Linear Algebra.)

COMPUTER SCIENCE

This is basically the course CS 1, Computer Programming I, described by the ACM Curriculum 78 Report [1]. The course introduces problem solving methods and algorithm development, teaches a high level programming language, and discusses how to design, code, debug, and document programs. I concur with the recommendation of the Computer Science Subpanel of the Tucker Panel [3] that at least a second course in Computer Programming is highly desirable.

References

[1] Association for Computing Machinery, "ACM Curriculum 78 Report," Communications of ACM, March 1979, 147-166.

[2] Bell, E. T., Men of Mathematics, Simon and Schuster, New York, 1937.

[3] Committee on the Undergraduate Program in Mathematics of the Mathematical Association of America, "Recommendations for a General Mathematical Sciences Program," Mathematical Association of America, 1981.

[4] Roberts, F. S., Discrete Mathematical Models, with Applications to Social, Biological, and Environmental Problems, Prentice-Hall, Englewood Cliffs, NJ, 1976.

DISCUSSION

Roberts: We've talked a lot about the problem of whether it's possible to have a unified course that introduces the student to both continuous and discrete ways of thinking and to let everybody take the same set of courses. I don't really think that the latter is possible but, on the other hand, I feel it's very important for the student not to be forced into making difficult decisions too early. So, I believe that the first course should, in some sense, be unified in order to give the student an overview that will help in making a decision on a major. After the first course, later courses in my proposal are fairly traditional and similar to those in the Tucker Mathematical Sciences proposal. Some of the sequencing of these courses, however, is different.

The first course, namely calculus with discrete topics in it, has the beginnings of a marriage between the discrete and the continuous. It spends a fair amount of time on sequences and relations and a fair bit on graphs and things like that. I think it's possible to begin on the very first day of a course like this by giving an example, probably discrete, which takes a real problem, translates it to a model, sets up a solvable mathematical problem which is solved by an algorithm, and then end with an unsolved problem to teach students that mathematics is a live thing and that they can be part of it. In the course proper you can begin with some elementary notions of sets leading to cartesian products which allow you to define relations. Graphs can be introduced as relations, emphasizing graphs as models and giving lots of applications of them. From that you can go into functions and you can also treat functions as relations. Then we get to specific functions: linear functions, how to graph them, polynomials, periodic functions, followed by special kinds of functions called sequences. But sequences would be defined not just by equations but by recurrences! We would introduce the idea of recurrences and maybe how to solve some of them by iteration. Limits would be introduced through sequences and would be used to define power functions so you could think of 2^x as the limit of sequences of 2 to rational powers. Then the course turns into a rather more standard calculus course: limits of functions, continuity, derivatives, curve sketching, max-min, related rates, things of that sort. I think there are possibilities here to tie in the discrete topics with the continuous, for example, some graph theory things that you can do with max-min. For instance, if you have a disconnected graph with n vertices and want to find out what is the largest number of edges you can have, you can look at that as a calculus problem where you have two sets and you want to put k vertices in one set and n-k in the other and you want to find out how many to put in each. And that turns out to be a simple problem of maximizing a function. The course, I think, would have to end with anti-derivatives and indefinite integrals and simple differential equations. There is probably not enough time for the definite integral.

Wilf: I think we're heading toward a situation that's very much in harmony with the times. Namely, that the two subjects won't quite be married but they'll be living together anyway.

Ralston: If, as seems common now, courses which once required calculus as a corequisite are now postponed one semester, then this course would serve as a good preparation to such courses. The second semester mathematics course could then be more directly supportive of other courses.

Maurer: One of the things this proposal suggests that we need to think about at more length is what subjects should only be briefly introduced in early mathematics courses and which need to have more than this. For example, is it satisfactory to engineers to have just an introduction to differential equations in the first two years and a full course later?

Tucker (Albert): There are two musts for engineers and physicists, exponential differential equations and simple harmonic motion. And I think it is very well established that these can be introduced along with the functions. It's a beautiful way of binding things together.

Steen: I'm concerned about the volume of material in your course. Each topic might be treated so quickly and superficially that there wouldn't be enough time to understand the techniques and do the drill and develop problem solving ability.

Roberts: Remember that we're leaving out a block of the usual stuff, namely definite integration and much of indefinite integration. Also, I'm hoping with Herb Wilf that symbolic systems will mean we don't need quite as much drill as in the past.

Dubinsky: The message still seems to be that introductory mathematics is calculus plus whatever else we might be able to squeeze in. That still seems like teaching elementary mathematics as if computer science didn't exist. I think the changes that have to be made to get a curriculum for computer scientists as well as physical scientists and engineers require broader, bigger jumps.

Roberts: I don't really disagree. I may have been too conservative and not ruthless enough. But all I wanted to do was to present the idea that is is possible to have a marriage.

Principles for a Lower-Division Discrete-Systems-Oriented Mathematics Sequence

Alan Tucker
State University of New York
at Stony Brook
Long Island, NY 11794

1. A Sequence of Five Courses

The following five courses are designed first of all with the needs of computer science students in mind. These students would form the core of the initial enrollment base for this sequence. Motivating examples, modern applications and algorithmic approaches should be used widely.

Course A. Algebraic Methods. Rates of growth; inequalities; sequences, their summation and limits; difference equations; elements of combinatorial enumeration; matrix algebra.

Course B. Foundations. Elements of propositional calculus; set theory and Boolean algebra; induction; recursion; number systems; graphs; algebras.

Course C. Calculus. See CUPM Mathematical Sciences Recommendations for first calculus course (Chapter II of Recommendations).

Course D. Probability and Statistics. See CUPM Mathematical Sciences Recommendations for prob/stat course (Chapter VI of Recommendations).

Course E. Advanced Discrete Methods. See CUPM Mathematical Sciences Recommendations for discrete methods course (Section I.11 of Recommendations).

Sequencing: Most students would take these five courses, one course per semester, in the order listed. Courses A, B and C do not have any inherent ordering. Course A, a mixture of the old-fashion College Algebra course and the newer Finite Mathematics course, is conceptually the easiest of the courses and is a bridge from current high school mathematics to college mathematics. So it is natural to take it in the first semester of freshman year. Courses A and B are counterparts to CS 1 and 2 in the ACM Curriculum 78 and should be taken in parallel to (or before) CS 1 and 2 in the freshman year. The calculus course, course C, can be taken any time in the first three semesters. Based the fact that calculus courses were more rigorous a generation ago when they were preceded by a year of college algebra (and analytic geometry and trigonometry), we are inclined to recommend that course C come in the third semester, or concurrently with course B in the second semester. Course D in probability and statistics has course C as a prerequisite. Course E should be to course A what advanced calculus is to introductory calculus. Many students should wait until the beginning of the junior year to take course E because of the greater problem-solving sophistication it requires.

Mathematics versus Computer Science: The overall goals of the preceding five courses should be to develop the mathematical skills and reasoning needed to study computer science and certain other disciplines and needed to analyze problems encountered in subsequent careers. These five courses should not teach computer science per se. While examples and exercises in these courses may involve applications to computer science, the role in computer science of these mathematical skills and methods of analysis should be left primarily to computer science courses. That is, we are not proposing an integrated mathematics-computer science sequence. The situation is akin to the calculus sequence for engineers or a statistics course for social scientists.

2. Practical Problems of a New Lower-Division Mathematics Sequence

One of the great practical problems in trying to implement any new lower-level mathematics sequence is that most departments feel well served by the traditional calculus sequence. As a consequence, a new lower-level sequence would be an alternative in addition to the calculus sequence. To have the prerequisites for science electives and to preserve the option of switching to another science/math major, a computer science student- the prime target for our new sequence- might feel compelled to take both sequences. Forcing computer science students to take two mathematics courses a semester in their first two years is unrealistic and undesirable. With additional courses in computer science, the students would be left with little space for college distribution requirements and for the opportunity to sample other potential areas of concentration.

If we reject the idea of a new sequence of different courses for computer science students, our sequence must be compatible with, or an adaption of, the calculus sequence. (By the calculus sequence, we mean a two-year sequence including a separate course in linear algebra.) The sequence proposed above has only one course in common with the calculus sequence. However,engineering and science students frequently take course D and, with the growth of digital systems, many engineering students now also take a course like course E.
In addition, course A is a mixture of the old higher algebra course and the new finite mathematics course.

Thus course B is the only really new mathematics course, i.e., not contained in 1970's CUPM course syllabi recommendations. But course B is basically the Discrete Structures course long recommended by the ACM for computer science students (it was Computer Science course C in ACM Curriculum 68 and is supposed to be taught by math departments as one of five required math courses in ACM Curriculum 78). In addition, much of the material in course B was contained in the type of Foundations of Mathematics course that was required a decade ago of mathematics majors at many colleges as a sophomore-level introduction to formal mathematics and proofs. A more advanced version of course B, titled Applied Algebra, is discussed at the end of Chapter I of the CUPM Recommendations for a General Mathematical Sciences Program.

In summary, this new sequence is compatible with current curricula and the sequence could be offered now at many colleges by modifying one or two existing courses. Students following this sequence are missing the second half of first year calculus, multivariate calculus, and (more mathematical) aspects of linear algebra (course A would have about six weeks of matrix algebra). Probably the most important gap with these missing courses is differential equations, which get about 20% of the time in the first three courses of the calculus-and- linear algebra sequence of the CUPM Mathematical Sciences Recommendations. One "catch up" junior level calculus course could cover most of the important material in these missing courses. Mathematical sciences students should also have a full course in linear algebra (which, with course A as a prerequisite, could go further into eigenvectors and other topics frequently skipped in the current sophomore linear algebra course). This is just a reordering of priorities since courses D and E, part of the CUPM recommended curriculum for mathematical sciences majors, are moved up before the second calculus course from their current junior-year status.

This sequence is still no substitute for science and engineering students; the junior year is too late for catching up on calculus. Engineering faculty are constantly complaining that four, or even five, semesters of calculus do not cover all the topics that they want their studens to know. They want partial differential equations with some Fourier transform methods and complex analysis; they want Stokes' Theorem, etc.

3. Three Pedagogical Themes for Computer Scientists in the Lower-Division Mathematics Sequence

There are three themes that seem to underlie virtually all mathematical analysis in computer science. They are: i) symbolic manipulation and reasoning; ii) problem-solving in situations of "organized complexity"; and iii) formal mathematical systems. Symbolic manipulation and reasoning is the essence of all sets of computer commands. The combinatorial complexity of computer science theories and of most applications programming sets computer science apart from the natural sciences. Computer science theory has been forced to take a formal systems approach to be able to make general statements about the behavior of various programs written in various programming languages employing various data structures and running on various computing machines.

All topics in the above five courses should be taught with one pedagogical eye always on these three themes of mathematical analysis. Most topics in these courses were chosen because they involve, explicitly and implicitly, these themes. The value of mathematics in any science lies more in disciplined analysis and abstract thinking than in particular theories and techniques. This situation is especially true in computer science. When computer science faculty complain to mathematics departments about the mathematical shortcomings of computer science majors, the complaints are typically about "general mathematical weaknesses", "a lack of mathematical maturity" or "inability to think abstractly".

Symbolic manipulation and reasoning begins in high school with algebra, where x's and y's abstract concrete numbers in algebraic equations and A's and B's represent lines in geometry. Course A extends this start with inequalities and limits and then "goes into higher dimensions" in matrix algebra where a symbol represents a row or an array of numbers. Course B in foundations uses non-numeric symbols in propositional calculus, Boolean algebra, and graphs. Course C extends numeric symbolism from operations on variables to operations on functions. Courses D and E have symbolic richness although this is not their primary purpose: D introduces "random symbols", that is random variables, and E discusses generating functions and recurrence relations.

Problem-solving was also started in secondary school (actually word problems begin in first grade). But the combinatorics in course A introduces a more difficult type of problem-solving that lacks an algebraic formalism. Course B introduces graph-theoretic problem-solving and logical formalism underlying much problem-solving in computer science. Course E builds greater mastery of problem-solving in situations of organized complexity. Random phenomena in probability problems (course D) add a new complication in problem-solving. Of course, all five courses should have weekly problem exercises.

Formal mathematical systems are introduced in many forms in course B on foundations. Calculus, by its very name, is an formal operational system. Probability theory has a formal structure. Course E can stress a systems point of view with several topics.

4. The Impact of This New Sequence on Non-Computer Science Students

This sequence has three major influences on the non-computer science student. First, it represents a movement away from calculus as the first collegiate mathematics course back toward the second year course, taught with more rigor than today, that calculus was in the first half of this century. A realistic compromise is probably starting calculus in the second semester. There was sentiment for such a move on the part of some members of the CUPM Panel on a General Mathematical Sciences Program, but the idea seemed too controversial in that context. For computer scientists, the issue is not so much getting better prepared for calculus as it is learning more relevant mathematics needed immediately in the first-year computer science courses. Recall that the pressure to bring calculus into the freshman year came from physics departments who needed calculus for their first-year physics courses. The physics students need to be given formulas for differentiation and integration of polynomials and trigonometric functions along with a brief discussion of the concepts of instantaneous slope and area under the curve-- this can be done at the start of a physics course. On the other hand, physics and engineering students would probably benefit from a more solid calculus sequence. There are a variety of courses that could precede calculus for these students, but one natural candidate is certainly course A. A corollary of this approach is that high schools might de-emphasize teaching calculus, which they tend to do poorly, and teach material in course A, which they should be able to do better.

The second influence concerns students who have a weak high school mathematics preparation and students who do not develop an interest in mathematics until their last two years of college. Currently students must progress through most of the calculus sequence before mathematical horizons begin to open. This time-consuming process discourages freshmen not prepared for calculus and upper-level students who do not have the time to go back through calculus to get at areas of mathematics of use to them. CUPM has on several occasions been asked to suggest an alternative to a year of precalculus work for poorly prepared freshmen. CUPM has never found an answer. The sequence presented here could be part of the answer. It will not be helpful to students with weak mathematical ability, but it may be highly appropriate for reasonably able students who, because of poor counseling, have only two years of high school mathematics. The CUPM Mathematical Sciences Recommendations propose (in section I.9) a new type of mathematics minor for students who do not realize the importance of mathematics for their future studies and careers until they have been in college for several years, e.g. psychology majors who upon seeing mathematical models and statistical methods in advanced psychology courses realize the usefulness of mathematics for graduate study (a math minor might also improve their chances of graduate school admission). This new minor incorporated much of the material in the five course sequence proposed above; it would typically involve only two semesters of calculus plus some linear algebra.

The final influence involves a broader intellectual issue that arises from the central role of calculus in collegiate mathematics. Many non-science/engineering majors now require calculus; of the total fall 1980 collegiate calculus enrollment of 520,000, over half was non-science students. Mathematical modeling, statistics and computing, what Stephen White calls the new liberal arts, are widely used in the social and behavioral sciences today, and calculus is but one of the mathematical subjects used in this modeling. Calculus tends to be required in social science curricula, as it is in the ACM Curriculum 78, primarily as a vehicle to develop mathematical reasoning. Social and behavioral scientists analyze systems of organized complexity, just as computer scientists do, although much of their analysis needs to be of a statistical nature. They need the problem-solving skills and experience with formal mathematical systems (but much less symbolic reasoning) that are central to computer science.

Social scientists years ago rejected the eighteenth-century goal of seeking social theories with a structure akin to Newtonian mechanics, but they still use the mathematics of Newtonian mechanics for the quantitative training of future social scientists. This is a mistake.

DISCUSSION

Tucker: The premises of my paper are that discrete mathematics should have a central role in the curriculum and that, to do calculus properly today, it should come later to avoid the trivializing which has become so common. My focus is on a program centered around a computer science constituency.

There are three very important themes in my program - symbolic reasoning, problem solving and formal systems with algorithms running through all the topics.

The discrete mathematics course is a reworking of a tried and true course, the old college algebra course (e.g. see Hall and Knight) which, though precalculus, was a lot more difficult than calculus. There is no text for this course currently. The texts that were developed for the ACM discrete structures course in Curriculum 68 were very hard to teach even to juniors and seniors.

Ralston: Your curriculum is aimed at computer scientists. But if two separate tracks are not on, how do you compromise this curriculum with that in your Mathematical Sciences curriculum?

Tucker: If you accept the premise of delaying the calculus, then I think there is no problem in either case. I am more comfortable with this idea now than I was when I was writing my paper for this meeting.

Greber: For engineers and physicists this doesn't work because they need the elements of calculus early even if the physics is delayed until the second semester.

The more conservative approach that Fred Roberts takes seems healthier. I want students to get finite mathematics and computing at an early stage as well as calculus.

Norman: Perhaps to solve the problem we are talking about, we could do a little calculus at the start very much motivated toward the examples and applications needed in physics and enginering and do the rigorous development later.

Wilf: A frequently heard criticism of mathematics curricula is that they keep emphasizing tools and more tools but that the students don't see where it's all leading for too long a time. I get the feeling that this is starting to happen in computer science; the gratification is being delayed longer and longer.

Tucker: I think there are lots of applications floating around here. And you frequently use them to motivate the topics.

Tucker (Albert): One of the things frequently mentioned in favor of teaching calculus is its relation to geometry. There is a strong tradition that goes back thousands of years that geometry is the best of mathematics. I think it is important, however, to distinguish between geometry as a formal system and the pictorial aspect of geometry. If I thought of combinatorial mathematics as largely symbolic, I wouldn't be the combinatorialist I am. It's the patterns, the pictorial aspects and the graphs that play a big part for me. Therefore, whatever discrete mathematics we try to put into the first two years, we should go out of our way to use pictorial material as much as possible. I think that will have a great deal to do with the success of introducing discrete mathematics into the curriculum.

Ralston: How broadly do you construe the pictorial idea? Computer scientists are less and less enamored of flow-charts but they do like to present their algorithms in structured forms which retain certain kinds of pictorial aspects.

Tucker (Albert): Yes, I would include this. Mental images can be formed of things that are purely symbolic. Emil Artin, an algebraist, used to say that he thought of a field as an onion and you peel it and there are the subfields.

STATISTICS IN THE TWO-YEAR CURRICULUM

Richard A. Alo'
University of Houston Downtown College
Houston, Texas 77002/USA.

As we have been discussing what should constitute a new sequence of courses for the first two-year curriculum, several questions arise as to the placing of topics from the areas of Introductory Probability/Statistics.

1. Should some topics from these areas be included in this new curriculum?

2. If so, the effectiveness as to how well they are taught is highly dependent on what topics outside of Probability/Statistics are included. Hence what are these topics?

3. If no topics from Probability/Statistics are included in this two-year curriculum, then what topics should be covered in the two-year curriculum to adequately prepare students to take a course in their third or fourth years.

Of course what is decided upon relative to Probability/Statistics as we have just mentioned is highly dependent on what our overall long range goals and immediate objectives are. The Mathematical Association of America's Committee on the Undergraduate Program in Mathematics and its Mathematical Sciences Panel in establishing its "Recommendations for a General Mathematical Sciences Program" had established as one of its overall goals that the emphasis of basic courses being designed should develop the mathematical skills and reasoning needed to study computer science, engineering, etc. and at the same time, develop the skills and reasoning needed to analyze problems encountered in quantitative careers that are a result of such preparation. It was felt that the development of rigorous mathematical reasoning and abstraction from the particular to the general were two themes that should unify our recommended curriculum. In addition, it was felt that first courses in a subject should be designed to appeal to as broad an audience as is academically reasonable. We felt that applications should be used to illustrate and motivate material in abstract and applied courses. Whereas, the development of most topics should involve an interplay of applications, mathematical problem-solving and theory.

The "Recommendations", therefore, concentrates its efforts on general curriculum themes and guiding pedagogical principles for a mathematical sciences major. The report refocuses the upper level courses on the traditional objectives of general training in mathematical reasoning and mastery of mathematical tools needed for a life long series of different jobs and continuing education. A combination of problem solving and abstract theory is emphasized to initially develop most topics. The latter implies that theory when introduced should be theory for a purpose, to simplify, to unify, and/or to explain questions of interest.

With its ties to engineering and decision sciences, computer science can no longer be categorized as a Mathematical Science. However, undergraduates in either area

(Computer Science or Mathematical Science) should have courses of substance in the other area--they are closely related and heavily dependent on each other.

In a similar fashion, it has become a misrepresentation of statistics to present it as essentially a subfield of mathematics even though it makes essential use of mathematical tools, especially probability theory. Statistics is the methodological field of science that deals with collecting, organizing, and summarizing data and drawing conclusions from it. As Chairman of the Statistics Subpanel for the CUPM "Recommendations" mentioned above, it was my privilege to express our belief that a first course should concentrate on data and on skills and mathematical tools motivated by the problems of collecting and analyzing data. This belief elaborates further upon the ideas of the first paragraph emphasizing a problem-solving approach.

Applied areas such as robustness, exploratory data analysis, and the use of computers have pushed a rapid growth in the field of Statistics. It appears apparent that some of this new knowledge should be included in a first course in Probability and Statistics.

Considering the "Recommendations" of the CUPM panel and its subpanel on Calculus for the first Calculus course, this would serve as a prerequisite for the one semester Introductory Probability/Statistics course recommended by the Subpanel on Probability/Statistics. Of course, we should emphasize that our recommendations did specify beyond the calculus prerequisite a level of mathematical maturity consistent with the intended and "hoped for" attainment by the student to ably handle the concepts. Since most Calculus courses 25 years or so ago were more rigorous than they are now, (at that time they were usually preceded in College by Trigonometry, College Algebra, and Analytic Geometry), it would seem reasonable to precede the first Calculus course by courses in Algebraic Methods and Discrete Structures (Discrete Structures as had been recommended by the Association for Computing Machinery for Computer Science students and as ACM 78 recommended that Mathematics Department teach). As a consequence of this preparation what was recommended by the Subpanel on Probability/Statistics as a joint/combined course in Probability/Statistics could be taken in the fourth semester. I have provided an outline of such a course below.

A Statistics course during the first two years of an undergraduate curriculum then should give students a representative introduction to both the data-oriented nature of statistics and the mathematical concepts underlying statistics. Considering the fact that the first two-year curriculum of a college program usually requires a student to take a breadth and selection of courses and considering the fact that enrollment data seem to indicate that most students take only a single course in this area, we propose a one-semester Probability and Statistics course. However, we feel that a two-semester sequence is by far more appropriate. We emphasize an exploratory data analysis approach as a means of creating the problem-solving motivating and instructional format described previously. As we have noted in our CUPM report, we do not feel that a course in "EDA" by itself is a suitable introduction to Statistics, nor should we replace for example, least squares regression by a more robust procedure in a first course. However, we do feel that new knowledge renders a course devoted solely to the theory of classical procedures out of date.

We emphasized EDA as being an approach to emphasize our motivating experience in teaching and learning. EDA is a way of thinking about data analysis and also of doing it. The underlying assumption of the approach is that the more one knows about the data, the more effectively data can be used to develop, test, and refine theory. Thus, this approach seeks to maximize what is learned from the data and emphasizes adherence to the two principles of skepticism and openness. We must be skeptical of measures which summarize data since they sometimes conceal or even misrepresent what may be the data's most informative aspects. At the same time, one must be open to unanticipated patterns in the data since they can be the most revealing outcomes of the analysis.

In several areas of the mathematical sciences, one finds probability as an essential tool. To the computer scientists, whose interests lie in the analysis of hardware and software performance and in the design and analysis of simulation models and experiments, as an example, the subject matter of continuous probability and statistics is necessary. Hence, we feel that it is not possible to compress a responsible introduction to probability and coverage of statistics into a single course. Therefore, we repeat our recommendation (as specified in (1)), that the probability topics be divided amongst the courses on probability and statistics, discrete methods, and modeling/operations research.

In addition, prerequisites for our course require some familiarity with computer programming so that instructors may utilize library routines or pre-written programs. We strongly recommend that these routines and programs be utilized.

OUTLINE OF THE COURSE

We repeat here, for reference, the recommendations of the CUPM Statistics Subpanel. This course is quite appropriate for our discussion.

I. DATA (about 2 weeks)

   A. Random sampling--Using a table of random digits; simple random samples, experience with sampling variability of sample proportions and means; stratified samples as a means of reducing variability.

   B. Experimental design--Why experiment; motivation for statistical design when field conditions for living subjects are present; the basic ideas of control and randomization (matching, blocking) to reduce variability.

II. ORGANIZING AND DESCRIBING DATA (about 2 weeks)

   A. Tables and graphs--Frequency tables and histograms; bivariate frequency tables and the misleading effects of too much aggregation; standard line and bar graphs and their abuses; box plots; spotting outliers in data.

   B. Univariate descriptive statistics--Mean, median and percentiles, variance and standard deviation; a few more robust statistics such as the trimmed mean.

C. Bivariate descriptive statistics--Correlation; fitting lines by least squares. If computer resources permit, least-square fitting need not be restricted to lines.

III. PROBABILITY (about 4 weeks)

A. General probability--Motivation; axioms and basic rules, independence.

B. Random variables--Univariate density and probability functions; moments; Law of Large Numbers.

C. Standard distributions--Binomial, Poisson, exponential, normal; Central Limit Theorem (without proof).

C. More experience with randomness--Use in computer simulation to illustrate Law of Large Numbers and Central Limit Theorem.

IV. STATISTICAL INFERENCE (about 6 weeks)

A. Statistics vs. probability--The idea of a sampling distribution; properties of a random sample, e.g., it is normal for normal populations; the Central Limit Theorem.

B. Tests of Significance-- Reasoning involved in alpha-level testing and use of P-values to assess evidence against a null hypothesis, cover one- and two-sample normal theory tests and (optional) chi-square tests for categorical data. Comment on robustness, checking assumptions, and the role of design (part I) in justifying assumptions.

C. Point estimation methods--Methods of moments; maximum likelihood; least squares; unbiasedness and consistency.

D. Confidence intervals--Importance of error estimate with point estimator; measure of size of effect in a test of significance.

E. Inference in simple linear regression.

We feel that a firm grasp of statistical reasoning is more important than coverage of additional specific procedures. Books like Box, Hunter and Hunter give much useful material on statistical reasoning.

When probability must be compressed in a unified course such as the above, the instructor must continually question the amount of probability needed for basic statistics. For example, moment gathering functions and continuous joint distributions would have to be omitted.

Freedman, Pisani, and Purves have good material showing students how to look at data and how to be aware of so-called pitfalls.

As we have mentioned previously, an important part of statistics is data collection. It justifies the assumptions made in analyzing data and it meets practical needs. Box, Hunter and Hunter have good material on this and on the motivational aspects directed towards probability and sampling distribution as provided by experience with variability.

A more recent text by Koopmans follows much of the concepts and ideas related her It is divided into three parts: Exploratory Statistics, Statistical Inference-Concepts and Tools, and Statistical Inference-Methodology. We have made use of it here in our one semester course and have found it to work well with the students enrolled who were computer science, engineering, mathematics, and science students. The book uses hardly any calculus and a few students have entered (with special permission) the course with no calculus background, even though it was a prerequisite.

The topics considered in our probability section do not depend on the calculus. We have recommended that the usual topics needed in probability be divided into three groups: one group involved with our recommended Probability/Statistics course, one in discrete methods, and one in modeling/operations research.

Basic combinatorial analysis, mathematical logic, discrete probability, limits and summation, graphs and trees as topics covered in a discrete mathematics course would provide appropriate and mathematical maturity background for a one-semester probability/statistics course. Definitely the ideas are appropriate to the development of statistics and do encompass topics that "statisticians" consider. On the other hand, one readily realizes how statistics has influenced the development of the above areas; for example, considerations in the design of experiments has led to the combinatorial/group theoretical work with block designs.

Consequently, we see a one-semester Probability/Statistics course as we describe it preceded by a discrete mathematics course as a strong alternative to the usual calculus-based sequence. Other probability topics requiring calculus-based knowledge can be covered as we have already indicated in other parts of the curriculum, once a calculus sequence has been taken.

As we alluded to in the previous program, we do see the need for two courses in the mathematical science curriculum which will enhance our Probability/Statistics course namely a discrete methods course and a stochastic modeling/operations research course. The latter should include at least conditional probability and several stage models, stochastic processes. (see CUPM recommendations).

Of course, it should be emphasized that Mathematical Science Majors still need a full Calculus sequence to be prepared for applied courses such as probability and statistical theory, applied statistics, or probability and stochastic processes--several second-level courses that we recommended.

Further in this pedagogical stream of thought we emphasize that the maturity sought involves "problem solving" capabilities and symbolic manipulation and reasoning. Much of what mathematics has to offer the sciences and business lies in clear analytical and abstract thinking, that is, both quantitative and qualitative reasoning--whether that science be computer science, economics, social sciences, etc. Definitely the Introductory Probability/Statistics course does have symbolic manipulation and reasoning through random variables and random phenomena in probability problems do enhance problem-solving capabilities. In addition probability theory with its formal structure does also give the student a look at what constitutes a formal mathematical system, so often only seen in a lower level course if it is a Foundations course involving for example, Boolean Algebra, Propositional Calculus, etc.

Consequently, the above tied in to a Discrete methods course could offer extensive symbolic manipulation, formalistic reasoning and overall systems points of view. These capabilities as provided here are not only important to computer science, mathematics, engineering, and the sciences, but also to the social and behavorial scientists, for example, who also analyze complex systems, but for the latter much more of the analysis is of a Statistical nature. In addition, they also need problem-solving skills and experience with formal systems.

REFERENCES

1. Recommendations for a General Mathematical Sciences Program, Committee on the Undergraduate Program in Mathematics of the Mathematical Association of America, MAA, 1981.
2. Koopmans, Lambert H, AN INTRODUCTION TO CONTEMPORARY STATISTICS, Duxbury Press, Boston, 1981.
3. Neter, John, Wasserman, William, and Whitmore, G.A., APPLIED STATISTICS, Allyn & Bacon, Boston, 1978.
4. Wonnacott, Thomas H. and Wonnacott, Ronald J., INTRODUCTORY STATISTICS, John Wiley & Sons, New York, 1977.
5. Box, George, Hunter, William, and Hunter, J. Stuart, STATISTICS FOR EXPERIMENTERS: AN INTRODUCTION TO DESIGN, DATA ANALYSIS, AND MODEL BUILDING, John Wiley & Sons, New York, 1978.
6. Burr, Irving W., APPLIED STATISTICAL METHODS, Academic Press, New York, 1973.
7. Freedman, David, Pisani, Robert, and Purves, Roger, STATISTICS, W.W. Norton & Co., New York, 1978.
8. Moore, David, STATISTICS, CONCEPTS AND CONTROVERSIES, W.H. Freeman, 1979.
9. Tanur, Judith, et al, STATISTICS: A GUIDE TO THE UNKNOWN, 2nd ed., Holden-Day, San Francisco, 1978.

DISCUSSION

Alo: Several questions arise related to probability-statistics and a new curriculum:

- Should some topics in these areas be included in a new curriculum?
- If so, what topics outside probability-statistics are needed to support teaching in these areas?
- If not, what topics should be included in the first two years to prepare students adequately to take a probability-statistics course?

My paper describes a one-semester course in probability and statistics which could be taken in the fourth semester, which builds upon earlier courses such as those recommended for the Mathematical Sciences program. I feel strongly, however, that a one-year sequence is appropriate for most students.

It should be noted that some of the content of this course could be included in a discrete mathematics course, things like combinatorial enumeration and discrete probability. In any case, the emphasis in this first probability-statistics course should be on exploratory data analysis and problem solving.

THE EFFECTS OF A NEW COLLEGE MATHEMATICS CURRICULUM ON HIGH SCHOOL MATHEMATICS

Stephen B. Maurer
Swarthmore College
Swarthmore PA 19081

Introduction

The title is too narrow by half. It suggests the influence will go all one way, with changes at the college level filtering down to the secondary. Perhaps in the long run that will be true, but in the short run, to the extent that the interface between the two levels is now carefully synchronized, what is presently taught at one level constrains what changes can be made unilaterally at the other. Even in the long run, if changes at the college level demand major changes at the secondary level for which secondary teachers are not ready because of training and philosophy, then changes at the secondary level will be very slow.

To summarize the arguments below: high schools are now teaching very little of the discrete mathematics material this conference proposes students should learn, so in that sense colleges have a free hand to teach this material in the first two undergraduate years and need not worry about what is done or not done, or done well or not done well, with this material in high schools. Every college professor has sometime thought, "I wish the high schools didn't teach calculus; the little bit the students learn just messes them up." Whether or not such thoughts are fair, there will be no cause for similar thoughts about discrete mathematics, at least in the short run. Professors can, indeed will have to, start teaching discrete mathematics from scratch. On the other hand, if the current calculus sequence is replaced by an integrated calculus and discrete sequence, as opposed to separate courses, then the whole current advanced placement program will be thrown in disarray, and both high school students and high school teachers will be very unhappy. If calculus were replaced entirely as a subject in the first two undergraduate years for most students taking math (but this is very unlikely), the consternation and disarray will be all the greater.

However, secondary teachers are aware that new sorts of mathematics and new applications have become important, and they are already hoping to include some of this in their curricula in the 1980's. In particular, computing and simple applications of elementary mathematics to social and management sciences are already coming in. This means there is an opportunity for more discrete mathematics to come into the secondary curriculum in time. To some extent, what is needed to lay the groundwork in high school for the later study of discrete mathematics (so that professors need not start entirely from scratch) is for schools to do a better job of the precalculus topics which students and teachers often rush through these days to get to calculus. However, it is also vital that these topics be taught with an algorithmic viewpoint and with somewhat revised emphasis; the need for this change, even the nature of this change, is hardly perceived at all yet at the secondary school level.

In the short run, then, we must be very careful about how we introduce a new college curriculum, lest we rend asunder the ties between high school and college mathematics. In the long run, however, there are excellent opportunities for making the two levels fit together at least as well as now, if changes are carefully thought out and energetically pursued.

The Short Run: How the High School Curriculum Constrains Collegiate Mathematics Change

What mathematics is taught in the high schools now? Pretty much the same things as 20 years ago, when I went to high school, and 10 years ago, when I taught high school. To be sure, there have been many innovations in between, but rightly or wrongly, most have fallen by the wayside. The standard program today (but hardly universal) for students who continue mathematics in college is:

- 9th Grade - algebra I, which goes through quadratic equations.
- 10th Grade - geometry, a fairly traditional Euclidean version, mostly in two dimensions and mostly with two-column proofs.
- 11th Grade - algebra II.
- 12th Grade - a precalculus course.

What is in this precalculus course? It varies a lot, but according to Martha Zelinka of the Weston, Massachusetts schools, it contains "everything the students should have learned before but didn't, and everything the teachers should have taught before but didn't." That is, in addition to a lot of review, it covers coordinate geometry, especially of conics, functions and relations; exponential, logarithmic and circular functions; perhaps also vector geometry; permutations, combinations and elementary probability; sequences and series; elementary theory of equations.

Able students generally take each of these courses one year earlier or double up one year, and take a calculus course in the senior year, usually equivalent to half a year of college calculus, but sometimes equivalent to a full year.

It is important to point out what is not included above. Algebra still consists mainly of learning certain methods of simplication which solve certain traditional problems. Little is taught about proof in algebra. In particular, the viewpoint is not emphasized enough that algebra is the means of manipulating expressions into whatever form (simpler or more complex) is needed for the desired conclusion. There is also little about algebraic structures. Also not taught is mathematical induction. Or if it is taught, it is taught with the same narrow set of examples which were used in the past. That is, students think that induction is a special method for proving summation formulas. They are given no idea that it is the fundamental method for proving just about everything in discrete mathematics.

When combinatorics is taught, it consists of traditional counting problems solvable with permutations and combinations; recursive methods, e.g., difference equations, are not taught. Nor is the algebraic method of generating functions. Not surprisingly, even the notation needed to talk about general combinatorial problems easily - set notation, Sigma and Pi notation (especially indexed over sets), iterated Sigma notation, etc. - is also not taught. This may seem like a small point; indeed, most discrete mathematics textbooks seem to assume, when they start using such notation, that students can pick it up immediately or else have always known it! In my experience this is emphatically not the case. In fact, one of the greatest stumbling blocks students have is the inability to read and write useful notation easily, and these are not skills that many students pick up quickly.

Continuing with things not taught: probability problems are still those which can be solved by dividing one counting problem by another. Even simple stochastic models are generally not discussed in high schools. Matrix algebra is not discussed. Nor are such useful and easily begun subjects as graph theory.

To be sure, there is an alternate course to calculus which is sometimes given in 12th grade and which does some of the things listed above as not taught. At one time there were high hopes for this course (in terms of how it would affect the interface between high school and college, and in terms of what a mathematically educated person would know), but today it is usually billed as a course for weaker accelerated students. Consequently it doesn't have much depth or much of the hoped for effect.

The one thing I have left out, and which is very different from 20 years ago, is computers. Most schools have something, if only a single microcomputer or a single time-sharing terminal. Many have much much more. Most able students have gotten on the computer (at home if not at school) and have written, no doubt, some nice programs. At many schools there is some effort to integrate the computer into courses - using it to compute areas under curves (in grade 9 as well as in calculus), to approximate roots of equations, and to do statistical analyses of actual and perhaps simulated data. More often, however, computers are used simply to provide graphics and play games, only some of which is educational. At any rate, as far as programming computers is concerned, they are used mostly for numerical algorithms, that is, they are simply regarded as big hand calculators. (The latter have appeared in the schools too, but with some debate.) The more general concept of computers as universal symbol manipulation machines, and all the interesting mathematical questions arising from this viewpoint, are not touched on. This fact is abetted by the language most often used in schools, BASIC. This is a wonderful language for learning how to run your first simple program without much fuss, and for doing small to medium numerical computations, but it's poor for fostering good, structured, algorithmic thinking, or for giving one access to the most powerful programming techniques and data structures. (Note: the latest version of BASIC, BASIC 7, is a structured, recursive language, and thus overcomes most of these objections. However, BASIC 7 is currently available only at the home base, Dartmouth, and it is not clear whether it will ever spread broadly. Even if it were instantly available in the schools, it is not clear that teachers would or could avail themselves of the new features and change the way they teach programming.)

In short, currently in the schools computers provide a teaching tool, but not a mode of thought and not an object of mathematical study.

As indicated before, both students and teachers see college-preparatory high school mathematics as heading towards, and for able students culminating in, calculus. Indeed, the main source of pride in many high school mathematics departments is how many students yearly take and succeed in the calculus offering. There is a standard measure of this success: how well they do on the College Board's Calculus Advanced Placement examinations. There are two well defined courses, Calculus AB and Calculus BC, and corresponding three-hour AB and BC examinations, given each May. The former course is usually considered the equivalent of one college semester, the latter two. These calculus examinations, first offered in the late 1950's, were taken in 1981 by about 33,000 students, which is estimated to be greater than 60% of all high school students taking an AP (i.e., college level) calculus course, but perhaps only 30% of all those taking some sort of calculus course at the secondary level. High schools take AP courses and examinations very seriously. Most high schols which have the trained staff to do it pattern their calculus offering(s) after the AP curriculum. (More students might take the AP examinations except for the cost and the fact that many colleges given credit on the basis of their own placement exams during orientation, or simply on the basis of high school transcripts; also, students who don't plan to take more mathematics in college often don't participate.)

That calculus should be the culmination of high school mathematics was a fine idea when the AP program began; the mathematical world at large thought so too. This conference is built on the premise that calculus is no longer the sole keystone of advanced mathematics. This premise has gained wide support at colleges and universities, and there is growing enthusiasm there for curricular change. However, at the high school level the old idea still holds sway, and is reinforced by the current Advanced Placement program. This is going to make the transition to a pluralistic view much harder, for both the colleges and the schools.

One should not think it is necessarily high school teachers who will balk at removing calculus from its pedestal. Perhaps students - having heard from their parents or older siblings that calculus is the "real math" - will be even more conservative. Here at Swarthmore, when we introduced an experimental freshman one-semester discrete mathematics course, we offered it the first time in the fall. But we could lure hardly any freshmen away from calculus; mostly upperclassmen signed up. So the second year we offered the course in the spring. This gives students a semester to get "tired" of calculus, and give us more time to advertise. The enrollment doubled - from 9 to 18 - still not much of a draw when 250 students take calculus yearly.

We can conclude that the study of calculus in the high schools is not going to go away over night. Even if the Advanced Placement examinations in calculus were suddenly terminated, this would only raise a big howl. Besides they aren't going to be terminated, and shouldn't be. After all, in the physical sciences, calculus is still the keystone.

Therefore, even if discrete mathematics is introduced at the freshman-sophomore college level, even if it is introduced as part of an integrated sequence, in the short run many of the best students will continue to enter college having taken calculus in high school. Consequently, there would still have to be regular calculus courses at colleges (at least 2nd and 3rd semester courses) for these same students to proceed with. If not, then we penalize our best students by making them repeat. And if these students finish calculus separately from discrete mathematics, then they must take discrete mathematics separately from calculus. Finally, if a sequence of parallel continuous and discrete courses must be given anyway for these students, why also give an integrated sequence, except perhaps as an experiment in a few schools of possible value for the long run?

The Long Run

In the long run, one can hope, secondary education will change in response to changes at the collegiate level rather than constraining such changes. Is this a reasonable hope in the case of discrete mathematics? I think so. First of all, secondary schools are always sensitive to what colleges want, or at least to what they require. Second, there is already the realization at the secondary level that changes are in order related to the changes this conference has in mind. There have been strong calls for action by secondary mathematics leaders. On the other hand, none of these calls speak exactly to the philosophy of the current conference. Specifically, the study of algorithms is not seen as the central glue of the called-for changes. Disseminating this philosophy of algorithmic centrality to the secondary level is the major additional step which must take place if the secondary curriculum and the proposed new collegiate curriculum are to dovetail well.

Current Proposals for Secondary Change. First, the National Council of Teachers of Mathematics has a report announcing recommendations for the 80's called "An Agenda for Action." Of its 8 summary recommendations, 3 bear on the goals of this conference.

#1. "Problem solving be the focus of school mathematics in the 1980's."

Here problem solving refers to much more than the traditional textbook problems. In particular, more realistic applications, involving a broader variety of methods, and applying to additional fields such as social sciences and business, are intended. So are problems which require computing devices as aids for solutions.

#3. "Mathematics programs take full advantage of the power of calculators and computers at all grade levels."

The sort of uses the report mentions specifically are anlysis of data, simulations, and use as an interactive aid in the exploration for patterns. It is also suggested that all students become computer-literate citizens and that computer-aided instruction can be helpful but cannot replace student-teacher and student-student interactions.

#6. "More mathematics study be required for all students and a flexible curriculum with a greater range of options be designed to accommodate the diverse needs of the student population."

One of the subrecommendations to this is 6.3: "Mathematics educators and college mathematicians should reevaluate the role of calculus in the differentiated mathematics program." Specifically, the report refers to the similar line of thinking stated in the MAA's PRIME 80 conference report. Recommendation II.1 of the MAA report says: "The MAA should undertake to describe and make recommendations on an alternative to the traditional algebra-calculus sequence as the starting point for college mathematics."

It should be noted that NCTM has also done a survey of mathematics teachers and educational administrators, the PRISM Project, to see what current and proposed mathematical activities have the support of these groups. In general, there is very strong support for the recommendations listed above. That is, it is not simply that leaders wish these changes; teachers and local school systems seem receptive. (On the other hand, declining enrollment will make #6 above, at least, hard to implement.)

Second, the College Board, through its in-progress Project EQuality, has been attempting to bring about in college preparatory secondary education a rededication to high standards in 6 central areas, mathematics being one. Both E and Q are capitalized to emphasize that both equality of opportunity and quality of offerings are the goal of the project. For mathematics specifically, an internal College Board report (currently undergoing extensive review), calls for an increase in both the minimum and the desired amount of mathematics students learn in high school. Students should learn about "Applications and Problem Solving", the "Language, Notation and Logical Structure" of mathematics, and "Computers and Statistics", as well as most of the traditional "Algebra, Geometry and Functions". Under logical structure is listed "appropriate experiences in pattern recognition, algorithm development, and inductive reasoning". Under computers are listed both computer literacy and computer programming ability.

Third, at the same time the College Board has announced a new Advanced Placement Examination in Computer Science, to be given for the first time in May, 1984. This decision was made after several years of discussion about what sort of new mathematics-related AP, if any, to give. The proposed syllabus for the associated course is quite ambitious. It will cover an "honest" full-year introductory course in computer science, as presently given at many universities, rather than the courses or experiences in computing which most high schools now make available to their students, and which used to be what students learned in introductory university computer courses. The subject matter will be as much nonnumeric as numeric, and the emphasis will be on how to think and write well algorithmically, using block-structured programs and appropriate data structures. Also, the course will require a sufficiently sophisticated language, specifically, Pascal.

These are the relevant current recommendations with which I am familiar. It should be noted that some high school teachers have very strong objections to Project EQuality and to the proposed AP Computer Science course. As for the former, they don't see how they are to fit in all the old material and all the new material. (The report does suggest that mathematics is important enough that perhaps students should be asked to take more than one mathematics course at a time at certain points. But what will teachers of other subjects say to that? A more plausible way to make room, though perhaps one no more likely to happen, is for the earlier topics in high school mathematics to be taught in grades 5-8, where currently not much at all is happening mathematically.) As for the new AP, practically no high school gives such a course now. In fact, few teach Pascal, and many will have difficulty making it available with their existing equipment. Moreover, not only do most high school teachers not know Pascal, they certainly do not know much about the difference between a computing and a computer science course, and why the change has been made at the collegiate level.

The high school reaction to the new AP course should be monitored closely. Although the teachers I have spoken to are somewhat taken aback by the proposed curriculum for the new course, because it is an AP program they are taking it seriously and are eager for their schools to prepare themselves to give such a course. (The College Board intends to provide assistance in establishing summer teacher training institutes to help prepare high school faculty to teach the course.) It will be several years before we know much about how things turn out, but a possible moral for the advocates of the discrete is: if you want to bring about a major change in what schools teach, arrange for a new AP on the material you want!

Let me summarize what I think will happen at the high school level, relative to discrete mathematics, on the basis of the broad support for the recommendations above. That is, I think these things will happen even if there are no further changes towards discrete mathematics at the college level.

At least some additional topics from discrete mathematics will slowly get into the high school curriculum, or get additional attention there. Statistics and finite probability definitely will. The language and use of computer algorithms definitely will. As algorithms are used more and more, mathematical issues related to algorithms will seem more and more natural. That is, the infiltration of further topics from discrete mathematics will become easier.

What Else is Needed. As stated earlier, there is a caveat to this optimism. For all the planned changes at the high school level, none of them are quite what this conference has in mind. What is missing is any emphasis, let alone elevation to status of central theme, of the "algorithmic way of life". That is, while the envisioned revised secondary curriculum involves doing algorithms, it does not involve either consciously applying an algorithmic viewpoint to one's entire mathematical training, or applying mathematical analysis to the study of algorithms. That is, I have not seen suggestions that teachers

1) describe traditional mathematics topics in algorithmic language;
2) present algorithms as a proof method, e.g., if you come up with an algorithm which stops only if it finds what you want, and you can prove it stops, then you have simultaneously proved the existence of the object and shown how to construct it;
3) discuss the correctness and efficiency of algorithms, i.e., show how to apply logic, induction and counting methods to verify algorithms and determine their complexity; or
4) present the modern precise idea of an algorithm, and some of its particular techniques such as recursion, as among the great ideas in human intellectual history.

In short, algorithmics is not seen as a major mode of thought around which to tie much of the mathematics one will teach or learn. Rather it is seen merely as a powerful but ancillary computational tool.

If the collegiate curriculum does change to give algorithmics this central role, then even with the changes which will occur at the secondary level anyway, there will still be a rather large gap between school and college mathematics unless the spirit of this change reaches the schools too. Right now, I see no signs that this spirit is even beginning to reach the schools. There is not even the awareness there of the existence of such a spirit at the college level.

It is important to understand why there is such a gap in perceptions. The first reason is: this perception of the centrality of algorithmics to mathematics is still quite a new and minority viewpoint at the college and university level. After all, most mathematicians alive today (including this writer) were brought up to believe that the quintessence of mathematical method and beauty is the existential proof. Now it is suggested we should change our esthetics: proofs involving algorithms are equally, if not more, beautiful and central. One can reject this suggestion by insisting that the esthetic of mathematics should not, or by definition cannot, change; that is, this difference in attitude towards proof is precisely the difference in philosophy which separates mathematicians from computer scientists. I don't hold this view - I think it confines mathematics rather than liberating it - but even if one does hold it, it is irrelevant at the high school level. High schools, like most small colleges, are not going to have computer science departments. It is mathematics teachers who will teach computer science, statistics, operations research, etc., to whatever extent it is decided these subjects are appropriate at the secondary level. Therefore, high school teachers must be imbued with this algorithmic viewpoint, regardless of whatever regrettable steps university mathematicians may take to shield themselves from it.

The second reason for the gap: topics covered in traditional secondary mathematics are not, at least on first analysis, very suitable for illustrating the algorithmic point of view. Too many questions can be answered by formulas. For instance, if secondary mathematics were less concerned with solving quadratic equations, for which there is a formula, and more concerned with finding shortest paths through networks, for which there is not, then it would be natural to introduce the idea that a problem is solved when it has a provably correct algorithm, and well-solved when it has a correct algorithm of low computational complexity, and go on to study the mathematics of algorithm verification and analysis. But the shortest path problem is not as central as the quadratic equation. Neither is any of the many other examples I can think of with algorithmic but not formulaic solutions. One still has to become facile at elementary algebra at least before one can do much else, and for most students absorbing even elementary algebra takes <u>lots</u> of time.

The reasons just given for this gap make it clear that the retraining problem for high school teachers is of paramount importance. If this were a period when a flood of new young teachers were entering the profession, already versed in algorithmics from their just-completed college or university training, then the problem might not be bad; but of course, exactly the opposite is the case. Fortunately for me, this thorny issue of retraining is the topic of another paper for this conference.

The comment above that students need time to absorb algebra illustrates another thorny issue already mentioned earlier: there isn't much slack time to play with in the secondary curriculum. Even if calculus were removed as the finale of school mathematics, that would still leave open only the senior year, and that only for the honors students.

If teaching were somehow improved (and students also improved!) so that all precalculus material was learned in the two years of algebra and one of geometry (recall the quote early in the paper), that would leave another year; but I see little hope of such an "improvement". Consequently, the only way to get the algorithmic view into the high school curriculum is by very careful planning of how to interlace this new viewpoint with the old material.

Answers to questions schools haven't yet asked. The questions are:

1) What might we suggest to schools to be the advanced placement mathematics course if they wish to give a new one instead of, or in addition to, calculus?

2) Given that algorithmic mathematics seems rather foreign to current secondary mathematics, but that planned changes at that level may lead to a change of perception on this, in what ways might parts of discrete mathematics and "prediscrete" mathematics be put into the earlier years of school mathematics?

I regard the second question as more important, because if collegiate mathematics changes, schools will come up with their own answers to 1) in any event; but interweaving preparatory discrete material into the earlier years is harder, and no significant changes along these lines will occur unless strong guidance is given.

So as for 2), the most important point is that the viewpoint expressed earlier - algorithms aren't needed for secondary mathematics because questions on this level are solved by formulas - is misleading. An explicit algorithmic approach has not been used at that level, but it could be. Take that quadratic formula. If you want a computer to evaluate it for you, you've got to write a program with a few branches, at least if you want to distinguish between single and multiple root cases, and if you have a machine which balks (like most) on being asked to take square roots of negative numbers. Granted, the complexity of this program is not such as to make computer scientists salivate, but we are talking about 9th graders without much computer experience yet. More important, even if you don't have your computer solve the quadratic for you, rather you use your hand calculator or even hand calculations, you can still think about the formula as a summary for an algorithm. Indeed, the same can be said of many procedures in secondary mathematics: solving triangles, graphing equations, solving simultaneous linear equations in two or three variables.

Again, an explicitly algorithmic approach to such topics in vacuo is unnecessary and maybe unnatural, but an algorithmic approach *is* necessary later on, in problems where formulas are not available. Why not start looking at problems this way sooner? At present, the change in approach a student confronts when he first reaches problems without formulas is more of a shock than it need be.

Here are a few more examples of topics which could easily be approached algorithmically. The best is Horner's method for evaluating polynomials; it is still taught in some places, if only in its equivalent form of synthetic division. There isn't any general formula for this in ordinary mathematical language (without using dot-dot-dot in a confusing way); all the books explain it by examples. But it has a very simple formulation as an algorithm. Also, it's clear that it works faster than "direct" evaluation, but it's not immediately obvious why what it outputs really is the value of the function. In short, this is one place in secondary mathematics where the issue of verification and analysis of complexity seem natural, and are not hard. Indeed, the whole subject of polynomials cries out for an algorithmic approach - to division, factoring, graphing, finding rational roots, approximating real roots, even approximating complex roots (an active research subject, but some basic algorithms could be studied by good high school students).

Of course, in one sense all of high school algebra is the study of solution "algorithms". I use quotes because the methods we use to solve harder problems in algebra are not all that well understood formally. We are just now getting "smart" symbolic manipulation software packages which seem to have taken algebra II and gotten at least a B. Getting students to formalize their algorithms for more than simple parts of algebra is too hard. But formalizing easier parts is well worth doing. It might even be worthwhile to go back and program the arithmetic algorithms of elementary school, e.g., multiplication and division of numbers in base 10 representation. The fact that such algorithms are taken for granted makes them no less brilliant or illustrative.

Another good topic for an algorithmic approach is sequences and series, which mostly means arithmetic and geometric series at present. Viewed algorithmically with recursive definitions, it is obvious that these are but special cases of linear recursions, and one might get much farther in the study of sequences in high school if one takes this recursive approach.

Of course, to the extent that computing is made an adjunct to the regular mathematics courses, there are lots of activities new to high school mathematics classes which can be introduced and which use algorithmic thinking. For instance, one can attack previously too difficult probability problems by number crunching or simulation, or look for patterns in elementary number theory, or give consumer math some content at last by doing realistically complicated financial programs.

But I am concentrating here on how standard topics can be viewed differently, not on how new applications can be added. Besides, examples of new applications with computers have been discussed for some time; see for instance [7].

To the extent that the training in computer use in schools becomes substantial, both in terms of sophistication of methods and in terms of size of problems handled, the mathematics of algorithms will become more natural. For instance, if recursive programming is introduced, it will not be obvious how many steps the programs take. Difference equations can come to the rescue and, incidentally, show that a recursive program is often longer to run than a corresponding longer-to-write iterative program. Another example: if students are told to alphabetize a long list of names using the computer, they may discover as they wait for their output that the usual first method, bubble sort, is not such a good idea; after that wait they may well be eager to see a mathematical analysis of the running time. Finally, if one learns some algorithms for which it is not obvious that they work at all, then verification becomes an issue. Fairly simple algorithms which come to mind in this category are: Euclid's algorithm; certain base change algorithms, especially for numbers with decimal parts; algorithms for certain solitaire games; the Gale-Shapley marriage algorithm; various graph search algorithms; algorithms for finding random combinatorial objects, say a permutation. But these rather quickly go far afield from concerns in school mathematics.

Finally, there are topics already taught in high school, but in light of their increased importance in algorithmics, not taught enough. Induction and counting stand out. They could be given increase emphasis, with examples drawn from algorithmic problems.

If the sorts of changes described above were made in school mathematics, then a college level course in discrete mathematics could proceed much more quickly than at present. The author's personal experience in teaching an algorithmic discrete course at the freshman-sophomore level is that I must spend a lot of time introducing induction, counting methods and algorithmic thinking, not to mention basic notation.

Now for question 1) about advanced placement. If discrete mathematics at the introductory undergraduate level is introduced and maintained as a separate course from calculus, then of course schools could simply offer both as AP courses. Given the NCTM recommendation that the school curriculum become broader and more flexible, schools would probably want to offer both. But if their resources are limited and they must choose one, as many will have to, which should we recommend?

Until recently, I would have said discrete mathematics. Like most mathematicians, I have felt intuitively that the discrete is intrinsically simpler than the continuous. Such a view, carried to its logical conclusion, underlies the ambitious Cambridge Report proposals of 1963 for how to change secondary mathematics. It was proposed therein, for example, that students should be introduced to difference equations in _junior_ high school, and that by the end of high school they should have covered more or less a complete course in difference equations paralleling in structure and sophistication the standard college course in differential equations. Such sophistication would be attainable, presumably, because the subject matter is finite.

But if discrete mathematics is simpler, then how come the sort of algorithmic discrete material we are talking about developed so much later than calculus in mathematical history? And how come discrete mathematics students find the subject harder than calculus?

The answer, I think, is this: because algorithmic discrete mathematics does _not_ have many formulas, or even many systematic methods for solving problems, students have no choice but to resort to "pure" mathematical reasoning. In short, the mathematics of algorithms requires considerable ingenuity and mathematical maturity. (If this is true, we have conversely that tackling such mathematics may be a good, if somewhat trying, way to develop such maturity.)

It is not that _doing_ algorithms is especially hard. Indeed, high school students seem to enjoy this a lot, and many are very good at it. But the _mathematics_ of algorithms is in parts quite hard. This should not be surprising. The "mathematics of calculus", i.e., the theory of functions of a real variable, is a lot harder than "calculus itself", i.e., learning the sorts of problems calculus can solve, learning to solve them, and developing an informal sense of why the methods are valid. We don't much try to teach the mathematics of calculus in calculus courses anymore, but leave it to the upperclass years. This is probably wise. The mathematics of algorithms is easier than the mathematics of calculus, but it may still be too hard for high school. Yet, if that mathematics is left out of a discrete mathematics course, it seems we are left merely with computing exercises and the jumble of watered-down topics in the old, finite mathematics course.

In short, if a school is to give just one AP mathematics course, I recommend calculus. This circumvents a lot of technical problems too, such as how are the high school teachers to be retrained for teaching a complete discrete mathematics course (perhaps an even more difficult retraining issue than how to prepare them to teach the algorithmic viewpoint throughout their courses).

What if the integrated two-year college sequence comes into being? Should a new AP course be developed which is half calculus, half discrete mathematics? If the colleges can agree on a fixed order in which to cover topics during the two-year course, so that schools could give a course covering the first semester, and an honors course covering the first year, then schools should do so; anything which allows able students to gain time at college, without skipping some things they need and covering other things twice, is to be applauded. Whether such a fixed order can be determined remains to be seen, especially since there doesn't seem to be any natural order in which to sequence things. To me, this seems like yet another reason to keep the continuous and the discrete in separate courses, at both the secondary and undergraduate levels.

Additions after the Conference

I see no need to delete anything from this article on the basis of the conference discussions, but I do feel that a few additions are in order.

The algorithmic way of life. Although many other papers referred to more or less the same list of discrete mathematics *topics* as necessary additions to the curriculum, nonetheless with the exception of Wilf's paper there was very little written support for the assumption I made that an algorithmic *frame of mind* should become pervasive. That is, there was very little mention that just doing algorithms isn't enough; rather, thinking mathematically in terms of algorithms should be the glue - also, there should be a new esthetic in which algorithmic proofs are better than existential proofs, and moreover the general circle of ideas surrounding induction and recursion are among the great ideas an educated person should know.

However, this viewpoint of mine was very definitely supported verbally at the conference. It is also stated explictly in the reports of the curriculum workshops which took place on the last day. These reports speak not only of topics but of overall themes, and framing one's mathematical thinking algorithmically is one of these themes.

Retraining. As emphasized before in this paper, taking such a pervasive algorithmic viewpoint in a new curriculum will make the retraining issue much thornier - even at the collegiate level. To be sure, mathematics professors already know the discrete mathematics topics proposed for inclusion, or could easily learn them, but to learn a new *attitude* is much harder. And for almost all mathematicians it will be a new attitude.

Moreover, if the retraining is going to be hard at the collegiate level, imagine the difficulty at the secondary level! This was brought home to me clearly when several of the very fine secondary teachers to whom I sent this paper told me that, when I got to talking about the algorithmic point of view, they really didn't know what I was talking about. In fact (as Dick Anderson pointed out at the conference), to many teachers the phrase "emphasizing algorithms" means "back to basics" - drilling in the classical numerical and algebraic manipulative skills so that students can do them without thinking, which usually results in students doing them mechanically rather than with understanding. Also not clear to my high school readers was the difference referred to between a computing and a computer science course. For instance, they were shocked to hear that computer scientists find something lacking in BASIC. Also, some of them asked me what Bubble Sort is; since sorting is an archtype problem for computer science issues, lack of familiarity with the Bubble algorithm (at least by that name) suggests lack of familiarity with these issues.

To be sure, I was writing for college and university mathematicians rather than secondary teachers, and I was aware that more detail would be necessary for the latter group. For instance, I didn't include any specific examples of what I mean by using algorithms to prove theorems, figuring my audience could supply several. (Here are the bare bones of one example. Consider the Euler circuit theorem, that a connected graph with an even number of edges incident at each vertex can have all its edges traversed in one continuous circuit. Avoid the standard proof by contradiction, which assumes the conclusion is false and considers a minimum counterexample G, and in which one shows that G would contain a cycle C and that G-C would be a smaller counterexample. Rather, give a precise statement of the "follow your nose" algorithm which finds a cycle in any graph G meeting the hypothesis, prove that it finds a cycle, and prove that after repeating this algorithm several times, the set of edges left in G must be empty. Finally, state an algorithm for merging these cycles, and show that the merger is a continuous circuit, as desired.) Even if I had stated several of my examples in detail, they are not about theorems which are well known to high school teachers.

The point is, even to introduce the issues of this new curriculum clearly to high school teachers will require a very different article than this one.

I had thought that the issue of how to retrain high school teachers would be discussed in the Weissglass paper, but all I knew then about the paper was the title, and I misinterpreted it. His paper was about how a new curriculum would be a boon in perspective to <u>future</u> teachers, whatever specific topics they ended up teaching, not about how to retrain current teachers to teach a new perspective. So let me say a little about this retraining issue now.

First, it really is a retraining issue. We can't expect there will be many new secondary teachers in the next several years. Second, there is an optimistic viewpoint about how current teachers will revise their perspective anyway, but I don't believe it. This optimistic argument goes: because computers are already in the schools, and will continue to play a larger and larger role there, the algorithmic point of view will naturally insinuate itself in time. That is, doing a lot of algorithms will eventually cause teachers to think about, and then teach, the mathematics of algorithms. I don't believe it because, if this were a natural and easy development, how come it hasn't happened already, pervasively, on the university level, where computers and large-scale problems which have to be solved algorithmically have been around longer and where faculty, in order to do research, keep abreast of changes? No, I am a pessimist on this: a lot of retraining of teachers will be necessary, and some of professors too.

How will this retraining be funded, in light of the current retrenchment? Fortunately, school systems and even teachers themselves are sometimes willing to pay part of the cost of, say, summer institutes. The teachers are willing to pay because their salaries go up if they take further courses or get further degrees. But usually the college or university running the institute is responsible for part of the costs and such institutions can only meet such costs from outside grants.

There has been a lot of discussion lately, at conferences of educators and in the press, about the dire state of math and science education in the United States. We can only hope that through concerted efforts such discussions will be translated into action at the highest levels, i.e., new government and foundation funding.

Is there room in the high school curriculum for changes? In the body of this paper I cautioned about thinking there was much room. I argued that a changed perspective on mathematics could be taught by carefully changing examples, but I warned against thinking lots of new topics could be added. Yet it has been suggested at this conference and elsewhere (see the Lucas paper) that there can't really be significant change in mathematics education until room is made for lots more material at the high school (and junior high school) level.

In his paper, Wilf argues that the $19.95 symbolic manipulation hand calculator soon to arrive will allow us to eliminate most of the drill from calculus, and thus considerably condense that course. In my paper, I do mention in the high school context the currently available, moderately powerful microcomputer forebears of such calculators, but I don't discuss whether the calculator-to-come could have a similar condensing role at that level. Much of high school mathematics deals with mastering symbolic manipulation; surely the ratio of skills to concepts is higher in high school algebra than in calculus. So, with that $19.95 machine, couldn't we condense high school mathematics drastically? What about with the $139.95 machine!

While I do think that the advent of this calculator (either one) will brighten the time-squeeze picture, I still caution against expecting too much. First of all, while there was considerable agreement at the conference that we should already do away with most arithmetic drill and allow students to use today's calculators instead (see Anderson's paper), in contrast there was considerable worry about what doing away with algebraic drill would lead to. Few think that skill in arithmetic is closely tied to general mathematical ability, but the relation between algebraic skill and mathematics is less clear. It may just be that skill and training in symbolic manipulation is closely tied to being successful as a mathematician, or scientist, or engineer, or even to being an astutely analytic businessman; if we don't "screen" students for other subjects by continuing to test if they have gotten good at traditional algebraic manipulation, we may not pick out those who are good at other sorts of symbolic manipulation either. Clearly much more needs to be known about this issue.

Another way it was put at the conference is as follows: We know what arithmetic skills we still want people to have in the computer age. They should know at least how to estimate effectively. That is, they still need good number sense. But what is the equivalent "good algebraic sense"? We don't know. Possible partial answers which were tendered: the ability to sense how many solutions a system of equations should have, and the ability to know what form an algebraic expression would best be worked into in order to draw from it easily whatever information is needed. But until we have a better idea of what algebraic sense people should have, we should be careful about throwing out the algebraic training we give now.

Let me say a bit more about this matter of converting algebraic expressions from one form to another. In the body of my paper, I say that schools spend too much time solving equations and not enough in more general manipulations from one form to another. It is the former skill, not the latter, which we can expect the Wilf calculators to perform. True, there is no reason to suppose that calculators won't eventually be able to do most of the algebraic rewriting that people have ever found useful, but it seems to me this is a lot farther off. Furthermore, the user will still have to tell the calculator what sort of rewriting he or she wants - at least if the calculator has not produced an appropriate rewrite on its own. In short, truly effective algebraic manipulation with calculators will have to be interactive.

This conclusion emphasizes for me the importance of continuing to give students thorough algebraic training. Without having had much training in arithmetic calculation, one can still know what sort of beast one wants when one has an arithmetic problem; but one can't know what sort of algebraic beast one wants without considerable training in algebraic calculation. Furthermore, as I point out in the paper, learning any mathematical skill well takes most students, even bright ones, a long time.

The advent of the Wilf calculator does, I think, present a wonderful opportunity nonetheless. It will naturally guide schools to spending more time on algebraic manipulation between forms, rather than in just getting the form x = some number.

One more point related to this issue of algebraic manipulation. It has been suggested to me that in this paper I continue the outworn tradition of giving the quadratic equation too much prominence. But I was quite careful. I did not say the quadratic equation was of crucial importance. I only said it was more important than the shortest path problem, a modest claim I think. But let me be more specific about how I think the quadratic equation continues to be important.

I agree that it is wrong today to make students spend lots of time solving the canonical quadratic over and over again - by factoring, by completing the square, and by the formula. But in fact, most teachers I know of don't make their students do a lot of this. The students spend much more time reducing some equation or equations in other forms to canonical quadratics, or taking a word problem and finding it is a quadratic, or maybe even in doing some more amorphous modeling problems which lead to quadratics. Now, of these 3 activities, the Wilf calculator may well do most of first (in addition, of course, to solving the quadratic), but it will not do the last two. Yet they are important, especially the modeling with quadratics (which admittedly, like modeling in general, is not done enough in the schools). Unless secondary students get experience with linear and quadratic modeling, then the only modeling experience they will get in school is modeling by computer simulation. One might argue that this is good, that almost all sufficiently accurate models are too complicated to handle analytically, and so the sooner students switch to computer modeling the better; in particular, one might claim that linear and quadratic models are almost always too crude to be useful.

Even if Newtonian physics didn't exist, I would reject this argument. With computer modeling only, one is apt to miss the forest for the trees. Put another way, because one is not forced to think through to conclusions with computer models, one is likely to miss patterns even when they are there.

Recently this was brought home to me forcefully in the course of some continuing work I am doing with an economist colleague. He brought me a certain model of economic dynamics. He had already successfully translated it from concepts into a set of differential equations. However, he had been unable to solve these analytically, either explicitly or qualitatively, and so he had already run many computer simulations. In particular, he wanted to know if a certain variable always changed monotonically. He believed it would, and it did in all his simulations, but he wanted to know if I could prove it.

After thinking about it for some time, it seemed to me (without yet a rigorous argument) that his belief was wrong, that if the initial values were related in a certain way, then the variable of concern would not behave monotonically. I told my colleague this. He doubted me, but was willing to run his program again with the initial conditions I proposed. The next day he phoned me to say that, upon looking at his previous reams of data again, he found he already had a case with that sort of initial conditions, and sure enough, the variable already had been non-monotonic - he had just not noticed it previously!

The story has a happy ending for my colleague. I went on to prove analytically that monotonicity wouldn't always occur, but in so doing I determined when it would occur, and this happened to include all the situations he was really interested in. But it's what happened in the middle that's important here. One cannot live by computer modeling alone, at least not without living dangerously! And if we don't teach some analytic modeling in high schools, most students will never get this message. Analytic modeling at that level will necessarily involve linear and quadratic models. Key properties of such models often hinge on the roots of a quadratic. Ergo, the quadratic equation is still needed!

Acknowledgments

I would like to thank the following people for helpful conversations and valuable written comments on an earlier draft: David Glatzer, Chancey Jones, David Levine, Dick Pieters, Tony Ralston, Cindy Schmalzreid, Irene Williams, and Martha Zelinka.

References

1. Advanced Placement Computer Science: Proposed Course Description, College Board internal document.
2. An Agenda for Action: Recommendations for School Mathematics of the 1980's, NCTM, Reston VA, 1980.
3. Goals for School Mathematics: The Report of the Cambridge Conference on School Mathematics, Houghton Mifflin, Boston, 1963.
4. PRIME 80: Proceedings of a Conference on Prospects in Mathematics Education in the 1980's, MAA, Washington DC, 1978.
5. Priorities in School Mathematics: Executive Summary of the PRISM Project, NCTM, Reston VA, 1981.
6. Project EQuality: Preferred Patterns of College Preparation, College Board internal document, 1982.
7. Recommendation Regarding Computers in High School Education, Conference Board of the Mathematical Sciences, Washington DC, 1972.

DISCUSSION

Maurer: If we go immediately to a rather radically different program with a completely integrated discrete mathematics and continuous mathematics curriculum or if we put off calculus for a year, we will very badly mess up the currently fairly good articulation between high schools and colleges. On the other hand, if we have separate courses and rearrange things more gradually, there will be time to work out a new articulation.

A peripheral point in my paper is that, if you really take the algorithmic point of view seriously, as I do, then it changes your whole idea of what you consider to be an interesting theorem or an interesting proof. You look for different sorts of proofs and you regard the importance of theorems differently. It's going to be hard to make the relevant changes at the high school level and even at the college level because few mathematicians are ready for this new view of mathematical esthetics. And without this new esthetics we are in danger of going back to the hodge-podge of topics that has been traditional in finite mathematics courses.

One thing that might be accomplished if we go to a program where calculus is not the entire focus of first year college mathematics is to loosen the hold of calculus on the high school curriculum. Many high schools now rush to get their students ready for high school calculus which may not be a good thing.

I've got mixed feelings about sweeping away a lot of the drill at the high school level using calculators and computers. I think it is important that students have some experience doing analytic modelling using linear and quadratic models without using machines. If you only work with the reams of data put out by the machines, you may miss the forest for the trees.

Tucker (Alan): There is a malaise at the high school level in that students don't find motivations for mathematics. This is one reason why so many stop mathematics early or come to college ill-prepared. One possible goal of doing finite mathematics is to try and enrich the high school experience by naturally tying the mathematics to computers.

Kreider: In arithmetic we need to do away with some of the highly specialized algorithms like long division while retaining the ability to do approximate computations used in the trial and error thinking which is the basis of inductive reasoning. For symbolic manipulation it is not so clear what can be given up and what should be retained to enable students to do general trial and error kinds of thinking.

Pollak: One thing that can be done away with because of even $4.95 calculators is the manipulation to reduce expressions to forms most easily calculated by hand. A lot of this leads to expressions which are in no significant sense "simpler" than the original.

Dubinsky: One of the things that happens in high school and college mathematics is that you never really get a chance to confront the concepts because the students get hung up on their inability to perform the algebraic manipulations. Although it's a very optimistic point of view, maybe using calculators to finesse the manipulation problem, we will be able to confront the concepts.

Lochhead: What is the possibility of starting discrete mathematics in the high schools as a way of avoiding the problem of starting calculus late in the colleges and the subsequent problems this would create for physics and engineering?

Maurer: Some of this could be done reasonably well and reasonably soon. And it's certainly true that, if the colleges started offering a finite mathematics course as their first course, than high schools would rush to offer such a course, too.

Young: I take a very pessimistic view of the high schools for the foreseeable future. The problem we're discussing is not just an American but a worldwide problem. In this country we're at the bottom of a very deep anti-intellectual hole and I don't see where the money - and it would take lots of money - is going to come from to pull us out. I think it would be very wrong to tamper the least bit with advanced placement calculus. That is one of the things that the high schools do do well. I also think it is utopian to talk about solving any of our university problems by moving any substantial body of material back to the high schools.

Barrett: I took part this spring in four locations in dialogues between high school teachers and college people on what kind of mathematics ought to be taught in the high schools. One motivation for the conferences I've been attending is that the demographics of the 1990s shows a shift in high school population with more inner city students. Right now the ACT scores from such students are very low. There is a real push to do something better for such students who typically take only two years of mathematics. Discrete mathematics may be a suitable change for students and one that ought to take place given the ferment in curriculum at the high school level.

Pollak: I find it difficult to accept Gail Young's notion that advanced placement calculus can't possibly change. Right now there is only one AP program so that's what you do. But if we can get agreement at the college level on discrete mathematics and computer science, then an AP course can be developed for those directions. Alternatives to high school mathematics don't get the current hearing they should now just because we have no AP possibilities.

Maurer: From discussions I've had at ETS, I think that, if there was a well-defined discrete mathematics course at the college level, it wouldn't be hard for them to develop a corresponding AP course. But if we do have an integrated calculus-discrete mathematics course and we can't get schools to decide what order to do the integration in, then I think we're in real trouble because the student won't know what AP course to take in order to fit best into college.

Lucas: One possibility, which is being considered in Israel, is to put some of high school mathematics into the junior high schools (which are very inefficient in the U.S.). Doing this might create an extra year in the high schools for finite mathematics.

Lochhead: At a recent convocation at the National Academy of Sciences on the problems of high school mathematics and science, one solution suggested was to have revolving positions between industry-like the computer industry-and the high schools. If this happened and high school mathematics teachers became good programmers, this would suggest new possibilities for discrete mathematics in the high schools.

THE EFFECTS OF A NEW DISCRETE MATHEMATICS CURRICULUM ON THE TRAINING OF TEACHERS OF MATHEMATICS

Julian Weissglass

Mathematics Department
University of California
Santa Barbara, California, 93106, USA.

This paper explores the possible effects on teacher education of including discrete mathematics (either as a separate course or integrated with calculus and linear algebra) in the lower division course work of mathematics students. The article is primarily conjectural. It is based upon my experiences with teachers and potential teachers, but it is conjectural nevertheless. In addition, even if I were an excellent judge of what the effects of introducing such a course might be, they could be totally eclipsed by the effects of technological, economic or political developments. In spite of these cautionary words, it is still worthwhile and appropriate for college and university mathematics teachers to think about how the inclusion of discrete mathematics in the college curriculum might affect precollege teachers.

Why should we be concerned with this at all? It is because of the rather obvious relationship between precollege and college educational programs. Each affects the other. In one direction, the kind of precollege education that students receive directly affects the colleges and universities implementation of their goals. After all, virtually every person who takes a college mathematics course, has experienced 10-12 years of school mathematics. Their knowledge of and, perhaps more importantly, their attitudes toward mathematics depends to a large extent on their previous teachers and courses. What we can do in 4 years hinges on how well they have been prepared by their elementary and secondary teachers. In the other direction, the primary means college and university teachers have for influencing the quality of our students' preparation is the contact we have with future mathematics teachers. The subjects we teach and the way we teach students will directly affect the teaching abilities of those who decide to become teachers.

Since secondary teachers would be more affected than elementary teachers by the proposed curricular changes, the bulk of this article will consider the training of secondary school teachers. Elementary teaching will be discussed at the end of the article. Henceforth when teacher is used without qualification it refers to a secondary level teachers (grades 7-12).

I shall consider the effects of a new curriculum on teacher training under four general categories: mathematical content, computing and algorithms, applications, pedagogy.

Mathematical Content

It is likely that at least some topics in discrete mathematics will be introduced into the secondary school curriculum during the working life of prospective teachers presently entering college. Moreover it is possible that substantial changes will be made in this direction. It is important that teachers be prepared now in these areas so they will be able to adapt to a changing curriculum. One of the major difficulties in achieving curricular reform in any area at any level is the need to prepare instructors to teach the new curriculum. Curricular reforms at the secondary and elementary levels often flounder because of this obstacle, since students who have recently studied the new concepts in college only slowly replace existing teachers. Attempts to retrain existing teachers are often frustrated by economic or emotional obstacles.

If we expect most teachers to incorporate topics from discrete mathematics into the secondary curriculum anytime during the next 30-40 years we must teach these topics to prospective teachers right now. The likelihood and value of discrete mathematics being introduced into the secondary curriculum is the subject of another paper at this conference. It seems likely that if changes are made they will come in this direction especially since the widespread availbility of microcomputers opens up the possibility of exploring areas of mathematics that previously could not have been done at the secondary level.

Moreover, even if topics from discrete mathematics do not filter down into the typical secondary curriculum, there is still good reason for having teachers learn discrete mathematics. After all, very few teachers teach calculus in high school and few of the other courses prospective teachers take (with the notable exception of geometry) are directly relevant to the secondary school curriculum. The justification for having prospective teachers study college mathematics is not that they will be teaching such subjects. Rather, it is expected that such course work will

(i) reinforce and deepen prospective teachers' understanding of mathematical concepts and skills,

(ii) provide perspective about mathematics as an intellectual discipline, and

(iii) increase their confidence in their ability to think mathematically.

How would introducing discrete mathematics into the lower division curriculum help fulfill these purposes?

(i) Certain areas of mathematics that are not reinforced by the calculus will be reinforced by learning discrete mathematics. These areas include proof by mathematical induction, counting arguments, properties of numbers, logic and probability theory. The strengthening of these areas would improve teaching abilities in first-year high school algebra and advanced algebra courses as well as in seventh and eight grade mathematics courses.

(ii) By learning a topic very different from the calculus the prospective teachers' perspective of mathematics will be changed. Many college mathematics students have the impression that mathematics is either a series of definitions, theorems and proofs or a rigid application of formulae. Calculus students in particular often feel that the answers are all known and there is a "right way" to get them. Topics in discrete mathematics (eg. graph theory or combinatorics) could be easily taught using an explorative or discovery approach which could give prospective teachers insight into how mathematics is actually done. Conjectures could be encouraged and unsolved problems presented. In addition, by seeing some algebra in an applied setting (for example Boolean algebra applied to switching circuits or semigroups applied to the modeling of sequential machines) prospective teachers would gain a view of algebra that other courses do not provide.

(iii) A course in discrete mathematics might well increase prospective teachers' confidence in their ability to think mathematically. The possibility of teaching such a course so as to increase problem solving skills would increase confidence in this area. Also devoting course time to a study of logic (the propositional and predicate calculus) would increase confidence in understanding statements of theorems as well as handling logical arguments in all areas of mathematics. I have found that undergraduate students at all levels have difficulty comprehending and using "necessary condition", "sufficient condition", "if-then arguments", "for every", "there exists", etc., and a study of logic would help remedy this situation.

Computers and Algorithms

Even a casual observer of the national scene during the 1980's will be aware of the impact of computer technology on our lives. The development of microcomputers (sometimes referred to as personal computers) has now brought this technology within the reach of many families and schoolrooms. In 1982 a quite powerful home computer costs under $400 and an extremely powerful one costs about $2,000. The mass production of microcomputers will decisively change our approach to mathematics education.

Many of the changes were foreseen in the early 70's, even before low cost microcomputers had become widely available. In 1975 the National Advisory Committee on Mathematical Education (NACOME) published an overview and analysis of school mathematics [4]. The report states:

> "The rapidly expanding availability of computers in schools and the equal rapid improvement in capabilities of the machines themselves mandates thorough reexamination of the content and methods of mathematics instruction at all grade levels. The nearly universal experience of mathematics teachers introducing their students to computing is a strong boost in student motivation and interest. There have been particularly noteworthy results with students identified as educationally disadvantaged. The demands of computer programming reinforce desirable mathematical methods such as organizing information, analyzing procedures systematically, checking answers for reasonableness, and finding errors. Furthermore, computer access increases student power to explore mathematical concepts by checking many examples in search of a pattern."

The report goes on to state:

> "Contact with computers must expand beyond a few very able mathematics students. The development of modules that integrate computing and regular mathematics courses is a sensible first step in realizing the extensive impact of technology on the methods of mathematics."

The NACOME report makes several specific recommendations.

In the area of curriculum content it recommends:

1) "that all students, not only able students, be afforded the opportunity to participate in computer science courses;

2) that school use of computers be exploited beyond the role of computer assisted instruction or computer management systems;

3) that 'computer literacy' courses involve students 'hands-on' experiences using computers."

In the area of teacher education it recommends that an emphasis be placed on seven areas, two of these areas specifically mention computers:

> "development of skills in teaching the effective use of computing and calculating machines in solving problems, for secondary teachers, literacy in at least one problem solving programming computer language and grasp of the issues in computer literacy."

In curricular development the NACOME report recommends revision or reorganization in the light of the increasing significance of computer and calculators.

Since mathematics teachers will probably be teaching about computers at the secondary level for the foreseeable future, it is important that they be comfortable with computers and competent to teach about them.

The issue of what to teach deserves additional consideration. Should mathematics teachers be teaching computer programming or should they be teaching how to use the computer as a tool in mathematics, or both. I think we will be selling secondary students short if we only prepare teachers to teach programming. Instead we should attempt to have teachers competent to teach their students how to use the computer in mathematics. This entails the following skills:

(i) The ability to develop an algorithm and write a program to solve a mathematical problem or to explore a mathematical situation, for example, to compute the greatest common divisor of two integers, to compute the representation of a number in a given base, to simulate a random process.

(ii) To use programs or packages developed by others, for example, statistical packages and software for graphing functions or solving equations.

(iii) To use "intelligent" computer programs to manipulate symbolic systems.

This last point is the subject of another paper at this conference, so will not be gone into in depth here. It should be obvious however that just as the advent of the handheld calculator resulted in educators questioning the amount of time spent in learning arithmetic skills, the capability of using microcomputer software that will allow the user to factor polynomials, do symbolic differentiation and integration and solve systems of equations will raise questions as to the amount of time spent teaching these skills. Will secondary school teaching change as a result, and if so, how soon? I think it is too early to predict, but it would certainly be advantageous for secondary teachers to have some familiarity with these new developments.

What should be our goals in the area of computer education of secondary mathematics teachers? I don't think it is realistic or even necessary to have all mathematics teachers become expert programmers. Remember that most teachers will be teaching mathematics, not computer science. However since it is likely that high school mathematics teachers will be teaching some computer science courses, they will need to know the fundamental concepts of computer science. As a minimum this would include understanding the capabilities and limitations of computers and knowing how to develop and encode a workable and understandable algorithm in a high level language. Their programs should be understandable, but not necessarily the most elegant or most efficient. A few extra seconds of running time is not crucial at this level. Most of all, prospective teachers should know how to integrate computers into the existing courses.

I think achieving these goals will require a minimum of two semester courses. One course could be a conventional college level course in a programming language. The second should be a course in "computer-based mathematics for secondary level teaching" or "the use of computers in secondary level mathematics". Certainly, the two courses could be integrated into a year long course. This would have the advantages of being sure that all students know the same programming language and integrating appropriate mathematical concepts completely with the introduction of programming tools.

The choice of a programming language will be controversial. Given the increasingly widespread availability of microcomputers and the prevalence of BASIC it makes sense for teachers to have some knowledge of that language. However, good arguments have been made for learning a structured programming language first. This would also be consistent with the goal of a discrete mathematics curriculum to teach an "algorithmic way of thinking" since structural languages are quite useful in helping one to think "algorithmically". Pascal is a structured language that has been implemented on some microcomputers. It is quite popular with computer scientists and would be a reasonable choice for a language. Recently, Logo, a derivative of the language Lisp, has been implemented on the Apple II and the Texas Instruments 99/4A microcomputers. Both Pascal and Logo are procedure-oriented which would help students learn how to develop algorithms. Logo has the advantage that it can be used interactively and is extensible (every procedure that is defined is treated as a primitive). In addition, Logo can be used by quite young children to explore the world of Turtle Geometry. Pascal has the advantage of more powerful data types and, since it is a compiled language, it runs faster.

Although an ideal teacher-training program would include a two-semester sequence designed specifically for prospective teachers, it seems unlikely that this is possible under present circumstances. A more realistic recommendation is that prospective teachers take the standard first course in computer science followed by a course in computer-based mathematics for secondary level teaching.

Since Pascal is commonly taught in the first computer science course, the second course could teach Logo or Basic or both. This second course should have as its goal showing teachers how the computer could be used to explore a particular area of mathematics or to solve problems. Particular emphasis would be given to developing algorithms. A wide range of topics could be explored. For example, the computer could be used to explore geometrical concepts. One way of doing this at a fairly advanced level is demonstrated in Turtle Geometry: The Computer as a Medium for Exploring Mathematics [1]. Other possible topics are numerical algorithms (roots, powers, logarithms, trigonometric functions, finding primes, computing the greatest common divisor, etc.), simulating random processes, finding solutions to polynomial equations and systems of linear equations, finding the area under a curve, etc. An outline of some topics suitable for such a course can be found in an article [3] by A. Engel. Such a course is necessary for the teachers to make the connection between their knowledge of programming and using the computer effectively in mathematics and in teaching mathematics.

Applications

The diversity of applications of discrete mathematics is addressed by several other papers at this conference. It is sufficient to point out here that knowledge of the applications of discrete mathematics will enable teachers to better communicate to their students the importance of mathematics in the world and about its diverse uses. Traditionally calculus courses have emphasized applications to the physical sciences. As a result secondary teachers have had limited knowledge of mathematical applications to the social and biological sciences. This could be corrected by a course in discrete mathematics and as this information filters down to secondary students, more of them might be encouraged to continue their mathematical studies.

Pedagogy

I stated at the beginning of this article that it would be mainly conjectural. One reason for this statement is that many of the effects mentioned above will depend on how the courses in discrete mathematics are taught. For example, students will not get an idea of the explorative nature of mathematics from discrete mathematics courses if these courses are taught using a definition-theorem-proof approach. Similarly, they will not become aware of the diversity of mathematical applications if few are provided.

The effect of discrete mathematics courses on prospective teachers teaching ability and methods is equally dependent on how the courses are taught. Some mathematicians will argue that this is not their concern; that mathematicians need only be concerned with mathematical content; that departments of education have the responsibility for teaching teachers how to teach. My position is different. People tend to teach in the way they have been taught. If students are passive listeners to lectures in their mathematics class, they will tend to lecture when they become teachers. If mathematicians expect mathematics teaching at the precollege level to be participatory, explorative and discovery-oriented they need to model it. About ten years ago, after one of my lectures, one of my students (a prospective teacher) said to me, "If you don't provide a model for how to teach mathematics to youngsters, what model will we have?" It was a poignant question.

In its recommendations for a general mathematical science program [2], the Committee on the Undergraduate Program in Mathematics (CUPM) states its program philosophy (p. 12). Several of the points deserve to be emphasized with regard to the effect of a discrete mathematics course on the pedagogical methods and skills of prospective teachers. CUPM writes that "The development of rigorous mathematical reasoning and abstraction from the particular to the general are two themes that should unify the curriculum". The theme of "abstraction from the particular to the general" is especially relevant to prospective teachers. They need to see a model of abstract mathematics being constructed step by step from concrete examples. Jean Piaget wrote that "Without a doubt it is necessary to reach abstraction ...but abstraction is only a sort of trickery and deflection of the mind if it doesn't constitute the crowning stage of a series of previously uninterrupted concrete actions". Often college level instructors assume that their students have had the concrete experiences necessary to understand abstract concepts. Acting on this assumption they fail to provide the particulars, the specific examples which are necessary to develop the abstraction. The result is that many students get overwhelmed by ideas being presented out of context. This affects their learning to be sure, but also affects the way they teach. They do the same thing to their students. Discrete mathematics courses, with their wealth of examples and algorithms provide an excellent opportunity to develop mathematical abstractions from the particular to the general.*

CUPM further recommends the use "of interactive classroom teaching to involve students actively in the development of new material" and that "applications should be used to illustrate and motivate material in abstract and applied courses." Both recommendations are particularly appropriate to an implementation of a discrete mathematics curriculum.

Discrete mathematics is well suited to an interactive approach. For many topics using a guided discovery approach or having the stu-

---

* It should be pointed out that for some students (those who have not had the proper concrete experience) even examples may not be enough. If the abstraction is to be understood, rather than rote-memorized, we will need to find creative ways of supplying the missing concrete experiences in ways appreciated by college age students. A full investigation of this point is beyond the scope of the present paper.

dents working in small groups with the instructor circulating among the groups (or some combination of the two) could be used. A small group approach would also be valuable in providing teachers with a model for using computers in a classroom when there is a limited number of computers available. Having the students work together in small groups to discuss concepts, solve problems, and develop algorithms while others are using the computer is an effective way of maximizing computer use. The advantages of a small group approach are described in articles by myself [6] and W. Page [5].

The importance of applications to school mathematics has long recognized. Discrete mathematics courses could play a useful role in providing secondary teachers with a wealth of real-world applications. They would then, hopefully, be able to answer the age-old question of the high school student, "What is this good for?"

Elementary teaching

Since most colleges and universities offer a special course designed for prospective elementary teachers and few elementary teachers take more than one year of college mathematics, the introduction of a discrete mathematics course will have mostly an indirect effect on the training of elementary teachers. However, for the relatively small number of students who study discrete mathematics and become elementary teachers the benefit would be similar to the ones described above for secondary teachers. In addition since many topics in discrete mathematics are closer in content to those topics taught in elementary school their understanding of elementary mathematics would be enhanced. Furthermore, the algorithmic approach would help teachers understand the development of arithmetical algorithms and to appreciate that there can be many different algorithms for a particular task.

An indirect effect of including discrete mathematics in the college curriculum is that it might accelerate the tendency to include suitable topics from discrete mathematics (e.g., graph theory, combinatorics, probability and statistics) in mathematics courses designed for elementary teachers. This is already happening to some extent.

Conclusion

The introduction of discrete mathematics course would have a beneficial effect on teacher training.

Teachers who learn discrete mathematics in college will

(i) be prepared for any changes in the secondary curriculum that would include topics from discrete mathematics,

(ii) deepen their understanding and perspective of modern mathematics and its applications,

(iii) develop additional confidence in their ability to think "mathematically".

To realize the full benefits for potential teachers discrete mathematics courses should be taught so that:

(i) it is interesting to prospective teachers;

(ii) it is presented at a level of abstraction they can comprehend;

(iii) it provides an enjoyable experience of exploring and discovering some beautiful and relevant mathematics rather than memorizing theorems and proofs;

(iv) it illustrates applications to real world problems.

Achieving these benefits will require that we challenge ourselves as teachers and mathematicians to develop new and appropriate curricular materials and to perhaps even learn to teach in new ways ourselves.

Acknowledgement

My appreciation to Phyllis Chinn and Neal Davidson for conversations about this topic and to Theresa Leibscher Weissglass for comments on the first draft.

References

1. Abelson, H. and Di Sessa, A. Turtle Geometry: The Computer as a Medium for Exploring Mathematics, The MIT Press, Cambridge, 1981.

2. Committee on the Undergraduate Program in Mathematics, Recommendations for a General Mathematical Sciences Program, Mathematical Association of America, 1981.

3. Engel, A., The role of algorithms and computers in teaching mathematics at school, in New Trends in Mathematics Teaching, Vol. 10, UNESCO, Paris, 1978.

4. National Advisory Committee on Mathematical Education, Overview and Analysis of School Mathematics, Grade K-12, Conference Board of the Mathematical Sciences, Washington, D.C., 1975.

5. Page, W., A Small Group Strategy for Enhanced Learning, The American Mathematical Monthly, 86 (1979), 856-858.

6. Weissglass, J., Small Groups: An Alternative to the Lecture Method, Two-Year College Mathematics Journal, 7 (1976), 15-20.

DISCUSSION

Weissglass: One question related to teacher training which I asked myself is why should teachers learn discrete mathematics at all. I think there are lots of reasons:

- it's inevitable that some discrete mathematics topics will filter down into the high school curriculum
- it would deepen their understanding of a lot of the things they teach in high school now
- it would give them a wider perspective on what mathematics is all about
- they will be better able to incorporate the computer as a tool in mathematics classes
- it will suggest many applications which could be explored in an informal or intuitive way.

In teacher training we need to go from the particular to the general and to do this by doing more than just giving an example and then proving an abstract theorem. We need to present mathematics as something creative and artistic and not as a rigid field of study to memorize.

All mathematics teachers should have at least an introductory course in computer science where they learn to program and the introductory concepts of the discipline. They need to see what kinds of mathematical topics are opened up by microcomputers and calculators. And instead of Basic, high school teachers should learn Pascal or perhaps even better, LOGO and the possibilities of turtle geometry.

Tucker (Alan): In Oregon some computer science courses designed by David Moursund are required of all mathematics teachers.

Steen: Although individual colleges may depart from the guidelines, it's the state requirements which are most important and these change very slowly.

Anderson: We're getting teachers in the classroom teaching mathematics today with very poor mathematics backgrounds. What comments do you have on this problem and the supply problem in general?

Weissglass: It is tragic that such poorly trained teachers are teaching mathematics. Summer institutes can help. I've seen amazingly dedicated, hard working teachers at such institutes. Although you have the danger that, with a little bit of knowledge about computers, teachers will just go into industry, microcomputers are attractive enough that they might attract people into teaching because of the possibilities for innovation.

Lucas: Another possibility besides institutes is sabbaticals to be used for retraining.

Weissglass: Another thing we must do is raise the image of high school teaching. Society must afford them professional standing. We put teachers in very difficult situations and expect them to do miracles and give them little appreciation and respect.

Bushaw: One way to enhance the standing of high school teachers is to be careful how we speak about them. Too often we're condescending even though there is a lot of good work being done out there as we see from the products. Indeed, sometimes, we lose mathematics majors in the colleges because they feel let down by the quality of instruction compared to what they had in high school.

Barrett: In addition to the status question, there is the problem of working conditions. With discipline problems even in suburban high schools and extra requirements without extra pay, it's hard to concentrate on education.

Lucas: One positive aspect now is the general recognition that we have a problem in the high schools. With AMS and MAA leadership we could establish useful contacts with high school teachers.

THE IMPACT OF A NEW CURRICULUM ON REMEDIAL MATHEMATICS

Donald J. Albers
Department of Mathematics and Computer Science
Menlo College
Menlo Park, CA 94025

1. What is Remedial Mathematics and Where is It?

"Remedial mathematics" refers to high-school level courses that are taught in colleges and universities, and includes courses in arithmetic, general mathematics, and intermediate algebra. Some universities and private colleges tend to regard courses in college algebra and trigonometry as being remedial.

From 1965 to 1980, remedial course enrollments nearly tripled in four-year colleges and universities, while total mathematics enrollments increased by only 52%. In 1965, remedial courses accounted for only 6% of total mathematics enrollments; in 1980, they made up 15% of the total. During the same time period, remedial courses also became more available among colleges and universities. Today nearly half of our universities offer intermediate algebra, and over half the public colleges offer elementary algebra. Very few private colleges offer remedial courses.

In two-year colleges, growth of remedial courses has been even more dramatic. From 1966 to 1980, they quadrupled and now account for nearly half of all enrollments. During the same period, total mathematics enrollments doubled. [1]

A host of reasons have been offered for the growth of remedial course enrollments, ranging from the influence of television to changes in family structure and open-door admission policies. One of the most important, and infrequently mentioned reasons, for their growth is the fact that in the 70's enrollments in high-school second-year algebra _decreased_ sharply, while total high-school enrollments _increased_. [2] Thus, _increases_ in remedial course enrollments seem to be due in large part to _decreases_ in enrollments of corresponding high-school courses!

2. Impact of a Redesign on Remediation

Redesign of the first two years of the college mathematics curriculum might have a wonderful impact on remedial mathematics, for at present remediation at the college level seems to mean teaching the same old high-school stuff--only LOUDER! A redesign incorporating some discrete topics carefully blended with calculators and microprocessors could go a long way toward increasing both the interest of students and teachers in that old high-school stuff. After examining remedial materials, one is drawn to the conclusion that remedial mathematics is remarkably dull.

It is generally agreed that teaching remedial courses is very difficult, so difficult, that many departments assign most of the remedial load to part-time faculty. A few departments have gone so far as to seek out remediation experts. These experts very often are patient, high-school teachers who are not bored by repeating the same material that their students previously "missed."

Remedial courses, as well as many service courses, seem to be the stepchildren of many departments and thus not watched over very carefully. In a perverse way, such a situation represents an opportunity for imaginative teachers to transform a dull rehash into an interesting experience for both teachers and students. It seems that at present there is far more opportunity for teachers at the college level to experiment and innovate in remedial courses than there is for teachers in high schools. By comparison, the high-school curriculum seems to be locked in concrete.

In spite of the "Back to Basics" movement, there are indications of some redesign of remedial courses and service courses. Textbooks now exist with titles such as _Arithmetic and Calculators_, in which calculators are meaningfully integrated into conceptual development. (One may also find an abundance of conventional textbooks with misleading titles such as _Arithmetic: A Calculator Approach_, which hold out promises of new approaches but, in fact, contain all of the old material with an additional chapter on calculators or additional exercises requiring use of calculators.)

In a few universities, some faculty are now integrating microcomputers into finite mathematics courses. They are providing students with software to handle various algorithmic exercises such as linear programming problems with more than three variables. They claim that this enables them to more fully discuss setting up word problems, to analyze the simplex algorithm instead of struggling with messy arithmetic, and to gain more insights into real-world modeling.

It would be especially interesting to see remedial courses leapfrogging more advanced courses through creative innovations. Realistically, this is not likely to occur if present staffing practices persist, i.e., if remedial courses are staffed primarily with faculty who may not feel free to modify their courses. Since not all remedial courses are taught by part-timers, particularly in two-year colleges, there is perhaps a better chance of revitalizing remedial courses in two-year colleges than in four-year colleges. This assumes little or no change in the content of calculus, which is now the "normal" first course. If redesign of the curriculum should result in discrete mathematics becoming the normal first course, then there is likely tc be considerably more pressure to modify present remedial courses.

3. _Limited Opportunities for Students Requiring Remediation_

There is hard evidence from Ohio State University that the student who takes these high-school level mathematics courses in college is severely limiting his/her vocational choices. In fact, the Ohio State University data shows that students who begin their college study at the arithmetic or elementary high-school algebra levels are much less likely to earn baccalaureate degrees than are students who begin their mathematics at precalculus or calculus levels. [3]

The surge in remedial mathematics enrollments has meant that large numbers of students have not been able to enter strong science courses as freshmen. Typically, students who do not study science as freshmen do not elect majors in science.

The data from Ohio State University shows that students who begin their study of mathematics below the level of precalculus (college algebra and trigonometry) have an extremely small chance of obtaining degrees in sciences or engineering. Moreover, students who start at the level of precalculus are much less likely to complete programs in science or engineering than are those who start at the calculus level.

4. What Mathematics for the Ill-Prepared Student?

Students Who Begin at the Arithmetic, or Elementary Algebra Level

If students who start their study of mathematics in college at or below the level of elementary algebra are not likely to end up in the physical sciences or engineering, then there may be a case for offering mathematics which better prepares them for careers in the trades, business, the arts, social sciences, agriculture, and elementary education.

Consider for a moment the usual content of arithmetic as recommended by the Mathematical Association of America and the National Council of Teachers of Mathematics.

Arithmetic

"The following basic arithmetical skills must first be mastered without the use of a calculator. These skills can then be extended and new mathematics learned by effective use of a calculator."

Computation with whole numbers, fractions, and decimals
Applications of percent
Translation of situations and verbal problems into mathematical statements
Facility in rounding and approximation
Understanding and use of basic arithmetic properties
Use and interpretation of graphs and tables
Computations with exponents and square roots

Most would agree that understanding basic arithmetic is highly desirable whether it be a student's last course or part of his preparation for advanced work. A basic problem with their recommendations for arithmetic is their insistence that these skills <u>first</u> be mastered without the use of a calculator. It would seem appropriate to see how the calculator might make the course more interesting and useful to students--to see which concepts can be better taught with the use of calculators (and microprocessors). Is it possible that arithmetic might be more interesting to adult students if linked to some ideas of data analysis? Is it possible that applications (or even drill and practice) involving the microprocessor will help develop organizational skills that may be of lasting value to students starting at this level?

Now let's examine their recommendation for elementary algebra.

<u>Elementary Algebra</u>

Arithmetic operations with literal symbols
Linear equations
Ratio, proportion, and variation
Operations with integer exponents
Operations with polynomials and rational expressions
Systems of linear equations with two unknowns
Special products and factoring
Solution of quadratic equations by factoring and formula
Use of formulas for area and perimeter of common geometric figures
Solution of elementary word problems

For many of these students, this may be one of their last mathematics courses, if not their last course.* Is it wise to offer them a review of elementary high-school algebra that is normally replete with phony word problems, simplification of complicated fractional expressions, and exercises on factoring of questionable value? Does not such a review help to convince them that mathematics is really not useful to them?

It seems that these students would profit from exposure to at least a few real-life applications facilitated with hands-on experience with microcomputers, visual displays, and some work with data analysis. Calculators and computers

---

*The students in this group are least likely to complete baccalaureate programs or even associate degree programs. It is estimated that only 20% of them complete degrees. Of those who do finish degrees, about half take degrees in the arts, education, and fields with no additional mathematics requirements. The remaining students in this group who complete degree programs seem to choose majors with little mathematics and perhaps take a course in intermediate algebra, finite mathematics, or statistics. Again, one can argue with little difficulty that these students would be well served with courses having more of an emphasis on topics from discrete mathematics and basic computing.

enable the student to spend more time thinking about approaches to problems. Such work can gently introduce important notions like flowcharting and algorithmic analysis. Graphic displays can help students go from graphs to algebraic expressions and vice-versa. Ideas from algebra that are needed for this work are more likely to be appreciated and remembered if their perceived applications value is raised.

Students Who Begin at the Intermediate Algebra Level

Next consider students who begin their study of mathematics in college at the intermediate algebra (second-year high-school algebra) level. As previously mentioned, students in this group are not likely to go into engineering or the physical sciences and have many similarities with the elementary algebra group. There is a suggestion that they have a stronger tendency toward business majors than do students in the elementary algebra group. Very often business curricula specify one-term courses in finite mathematics, soft calculus, and statistics, with prerequisites of intermediate algebra or college algebra.

The usual content of intermediate algebra courses as recommended by the Mathematical Association of America and the National Council of Teachers of Mathematics is as follows:

Intermediate Algebra

Simplification of algebraic expressions
Fractional exponents and radicals
Absolute value and inequalities
Operations on polynomials
Quadratic equations; completion of the square, quadratic formula, properties of roots
Quadratic inequalities
Graphing linear and quadratic functions and inequalities, determination and interpretation of slopes
Solutions of equations with rational expressions
Systems of linear equations with two and three unknowns; homogeneous, dependent, and inconsistent systems
Polynomial equations
Binomial theorem
Arithmetic and geometric sequences, infinite geometric progressions
Exponential and logarithmic functions and equations
The function concept, including compositions and inverse functions, arithmetic operations on functions
Analysis and solutions of word problems (including estimation and approximation) shall be emphasized throughout the above sequence of topics

Which of these topics would you willingly delete or reduce in scope in order to introduce some discrete topics if the goal is finite mathematics or soft calculus?

What if the goal is a discrete mathematics course? Probably most would argue that most of these topics are important to both work in calculus and discrete mathematics.

It does seem, however, that many of these topics are dull standing alone and suffer from lack of application. They might very well be enhanced by integrating algebra topics with counting problems, organizational problems, and computing fundamentals. Certainly, some of the topics might be made more interesting through the injection of the following notions:

1. Quadratic inequalities could be enhanced by applying them to linear programming problems of two and three variables.
2. Linear programming also can provide some nice examples of word problems. Using a prepared linear programming algorithm could be nicely linked to the computer and graphic displays.
3. The solution of equations could be made more interesting by examining elementary numerical methods for approximating solutions via calculators and/or computers.
4. The calculator can also be used to great advantage when working with exponential and logarithmic functions.
5. Calculators can be a valuable assist in graphing.
6. Calculators (i.e. function machines) can be used to drive home the notion of function and inverse functions.
7. Certainly, tables should <u>not</u> be used unless absolutely necessary.

Note that I suggested hands-on experience with microcomputers and graphic displays in both elementary algebra and intermediate algebra. There are two basic reasons for the suggestion. First and foremost, they can be used meaningfully to motivate old, but important ideas of algebra with applications closer to the experience of students at this level. Second, students <u>know</u> that microprocessors are important and useful and that they often deal with numbers, i.e., with mathematics. To ignore cheap, modern computers is to enforce the suspicion of students that their mathematical skills are not useful. Thus, it is very important to earnestly seek "honest" applications at this level that are easily handled by computers, and to thoroughly examine all standard topics in terms of enhancement via computer links. Given current student interest in computing, it is in our best interests to intelligently utilize them in mathematics and to ask how current content might be enhanced through their utilization. In view of the competition for time within the curriculum, and the clear rise of computing with its strong links to mathematics, we must face the fact that some parts of the curriculum will be reduced in order to let computing in.

5. Remedial Discrete Mathematics?

What can be said about students who begin their study of college mathematics at the precalculus level? As previously noted, students in this group are less likely to complete baccalaureate degrees in the physical sciences or engineering than are students who begin at the calculus level. It is estimated that up to half of the students in this group end up in the sciences, engineering, and business. Greater numbers of this group of students might very well be nudged toward careers in computer science if the precalculus course were modified to become a prediscrete mathematics course.

First, most if not all of the trigonometry, could be eliminated. Work on topics such as coordinate geometry and polar coordinates is hard to justify if the next course is to be discrete mathematics. This would provide considerable time to add interesting and lively topics that would help students bridge the gap to a regular discrete mathematics course. These topics might include rudimentary combinatorial analysis, flowcharting, finite probability, the discrete number system, critical path analysis, and elementary analysis of algorithms. All of these topics possess real-life applications that can be naturally tied to the computer in a very meaningful fashion. [3]

It is clear then that the gap between intermediate algebra and discrete mathematics is much less than the gap between intermediate algebra and calculus.

There is a danger that some will regard the gap between intermediate algebra and discrete mathematics as being so small that they might consider going directly to discrete mathematics. The experience of many teachers is that most students find their first exposure to ideas from combinatorial analysis and discrete number systems difficult, primarily because they have had no previous experience with those ideas. To that end, let us develop a prediscrete course, chock full of examples, and wed it to the microprocessor.

References:

1. J. Fey, D. J. Albers, and W. H. Fleming. Undergraduate Mathematical Sciences in Universities, Four-Year Colleges, and Two-Year Colleges, 1980-81, Washington, D.C.: Conference Board of the Mathematical Sciences.

2. Joan Leitzel, "The Developing Crisis in the Mathematics Classroom: Causes and Some Cures," Summary of presentation at AAAS annual meeting in Toronto, 1982.

3. J. Whitesitt, "Mathematics for the Average College-Bound Student," Mathematics Teacher (February 1982): 105-108.

4. B. H. Burkhardt, The Real World and Mathematics, Blackie, 1981.

5. Z. Usiskin, "What Should Not Be in the Algebra and Geometry Curricula of the Average College-Bound Student?" Mathematics Teacher, (September 1980): 413-424.

DISCUSSION

(Due to illness Professor Albers was unable to attend the conference)

Ralston: Is it true that almost all the students who come to college needing remedial mathematics are not going to study scientific or technical disciplines? And, if so, is the whole remediation question somewhat irrelevant to our concerns at this meeting? Or, even if the numbers of such students relevant to our concerns is small, is there reason to pay special attention to them?

Lochhead: Remediation is certainly not irrelevant to us if we are concerned about getting women and minority groups into science and technology.

Young: Still, it is true that those taking remedial mathematics are not going into physical sciences or engineering. My data suggest that you won't make it in these disciplines without four years of high school mathematics.

Barrett: But we can't ignore the increasing percentage of high school students who are going to be coming from lousy urban school systems. We need programs like Project Upward Bound to give these students a chance at scientific and technical careers.

Anderson: Although the data is confusing, there is some information that the number of people taking three years of high school mathematics is greater than it was a few years ago. Partly this may be that more women are taking mathematics in high school.

Maurer: There are two things that should make us keep in mind the remedial issue:

1. More students are coming into the colleges and junior colleges who have had little mathematics in high school but want to take mathematics in college. At the junior colleges, in particular, we might try to get the faculty to teach mathematics to such students with an algorithmic and discrete flavor.
2. Although not quite a remediation issue, there are students even in the best colleges and universities, often minority students, who have had three or four years of high school mathematics but have a rough time in college mathematics. We need to pay attention to them when thinking, for example, about honors, regular and slow versions of a new curriculum.

Pollak: A few years ago there was a survey done about the continuing education desires and needs of two-year college faculty in mathematics. As I recall, the two main things that the faculty themselves wanted was additional knowledge in applications of mathematics and in computing.

Barrett: But since the majority of community college students are not taking pre-baccalaureate courses, we must understand that mathematics in the community colleges is a whole different breed of cat from what our discussions have primarily been about.

Dubinsky: I want to throw out a suggestion that might be useful in teaching remedial mathematics and also in teacher training programs. It may or may not work better if you first teach students to program in a language they can learn very quickly, like Basic. But then what you do is to teach students to program in a very high level language, for instance SETL which is a very mathematically oriented language. In order for the students to learn to write correct programs in this language they must learn many concepts in discrete mathematics because that's what the language is all about. If you ask someone to write a simple program in a language on a terminal or whatever, that gets to be seductive. They start to get involved. They're not thinking about learning discrete mathematics; they're thinking about getting the correct program.

Lochhead: That's the philosophy of LOGO with respect to geometry.

FINITE MATHEMATICS - THEN AND NOW

John G. Kemeny
Dartmouth College
Hanover, NH 03755

When you are invited to give an after dinner speech I suppose one should tell an after dinner story. I will do that by recounting a story that happened more than a quarter of a century ago. It is the story of an organization called the Mathematical Association of America and a committee of that organization, whose initials in those days were CUP though somehow later it sprung an additional initial of M.

The purpose of that organization was to change the first two years of college mathematics. Specifically what they were most interested in was to break out of the lockstep that was then traditional, that everybody studied calculus for the first two years. They tried to bring about a curriculum change in which mathematicians would recognize that that no longer made sense, but rather that there should be some room for discrete mathematics in the first two-years curriculum and that a broader range of mathematical ideas should be exposed to all students. What they came up with was a Freshman year which was half basic ideas of Calculus and half discrete mathematics; in the sophomore year they recommended divergence depending on whether the student was going into the physical sciences or the social sciences. But even for the sophomore year the committee recommended a quite different curriculum for those who were not going into the physical sciences, which is quite an interesting course. In addition to that the committee developed some experimental text materials to try to show what these courses might look like.

This happened at a time I was a very young chairman of the Mathematics Department at Dartmouth College. I have to confess to you that in the beginning I had never heard of CUP. As a matter of fact, as one person in the room knows, I had not even heard of the Mathematical Association of America. Having been a good Princeton Ph.D., I naturally joined the American Mathematical Society but I did not know that the other group was around until a person by the name of A.W. Tucker came to visit Dartmouth College to see what some of his ex-students were doing in that far away and strange place, and discovered that what we were engaged in was implementing the recommendation of a Dean.

It was a quite remarkable Dean at Dartmouth; he was a political scientist and he wasn't bothered by what mathematicians were teaching but by what mathematicians were _not_ teaching. (This was in the mid-50's.) He was sure mathematics would become very important in the future of social scientists and he did not see anything in the mathematics curriculum that was suitable material for social scientists. He wasn't sure what kind of mathematics it should be, but he asked if we would be willing to undertake such a project, and was even willing to give us money to get some very distinguished social scientists to come to campus to teach us a bit about what social scientists do. None of us knew anything about that strange area, but three pure mathematicians said "OK, we're willing to learn - let's try it."

We developed a course that was entitled, "Mathematics for the Social Scientist", and we offered it as an experimental section getting permission to try it once. When we tried to re-offer it to a limited group, 165 students signed up and we had to go to an emergency meeting of the science division to get permission to change it to a regular course. At that division meeting we almost lost the fight. The science division liked the course, but one of the senior faculty members said that the title was unacceptable, because you should never title a course by the clientele; it should describe what the subject matter is. And we didn't know how to describe what the subject matter was! On the spur of the moment I thought of a title that might be acceptable. I told them that I was inventing a phrase, there was no such phrase, but the division bought it. An editor who later became a good friend of mine told me (years later) that when they first heard about our book, he was willing to take a chance on a textbook for a course that was only taught at one college in the entire United States, but he told the salesman he would _never_ publish it under that title, because nobody would ever buy a book called "Introduction to Finite Mathematics".

The book was published, as you know, in spite of his prediction, and both it and the 50 or so successors to finite mathematics sold very well. If you think back, as Tony Ralston asked me to do, as to what the various finite math books achieved, there were two places where they made a major impact. The biggest one was as an alternative to the Calculus for the students who wished to take an alternative. (If you look at Gail Young's article for this conference, the statistics show that of college level courses, next to calculus, finite mathematics was by far most heavily elected.) Secondly, there was a use of it that we did not foresee at all! At many small colleges the finite math course was the first course that had any Twentieth Century ideas in it.

Let me now say what it did not achieve. Thanks to Al Tucker I did join CUP. I became completely convinced of the eventual recommendations of the CUP, namely, that the change should be for everyone and that two years of calculus just did not make sense in 1956. I believed it then, and I believe it now. As a matter of fact, very briefly we achieved that at Dartmouth. There was a period when all students were forced to take something other than calculus, if they took two years of mathematics. But even at Dartmouth that battle was eventually lost when Physics and Engineering lobbied back and cut the version of finite math that was designed for that sequence out of the curriculum. So finite math achieved a number of things that I hope were worthwhile, but it never achieved what I had hoped the major role of it would be.

If you tell stories, there should be morals to stories. The moral of that story is that the CUP was absolutely right but it did not achieve its goal. Its recommendations were not carried out, the experimental textbooks (and I am co-author of three out of four of them) were not used or were hardly used at all. Bob Norman, incidentally, was the editor of the third and fourth, which were written at Dartmouth and were of excellent quality. The important impact was made by commercial textbooks. I think that is a lesson this group must keep in mind: that committee recommendations may be important in terms of what it is worth doing, and experimental textbooks may get you started, but unless you get some major commercial publisher to push your wares, you are not going to succeed!

I was asked to speak about the role of discrete mathematics, then, and more importantly, now and in the future. I will do that, but I decided first to speak on something else, because to concentrate just on discrete mathematics is a mistake. I would like to argue first for some changes in the way <u>all</u> mathematics is taught and therefore I am going to use as my example freshman calculus.

Since I was reminiscing, I thought that I would wake you up after dinner by saying: "I taught my first calculus course 38 years ago." I just turned 56 and I think you can do that arithmetic. I was eighteen and everybody in my class was older than I was. That's a long time ago. I've taught calculus a great many times in 38 years, and I have to tell you that I teach a very different course today from what I taught 38 years ago. The reason is not only more experience, though that does contribute, one does improve with age, but I've changed my mind about what is important in calculus and I changed it quite fundamentally. And I've changed it sufficiently that I am very upset by the fact that most mathematicians have not yet changed <u>their</u> minds.

Just to give you a quick example, I believe that the typical homework assignment one gives in techniques of integration constitutes cruel and unusual punishment, and has a great deal to do with the fact that many students hate and fear calculus. If there ever was a good reason for teaching that, those reasons have totally disappeared.

But let me go a step further than that; it's too easy to make that kind of point. Let me tell you something deeper that I thought about a good deal in the last couple of years. I think mathematicians have tremendous pre-occupation with formulas, because they love finding answers "in closed form". Consider the simplest kind of integration problem, say

$$\int_0^{13} e^x \, dx$$

It's the kind of problem you put on a freshman calculus exam so that every student should get at least one problem right. Everybody knows how to do it. The answer is $e^{13}-1$. Suppose someone came along and said, "Why didn't you use numerical integration?" You'll say that is absolutely mad. Here is a closed form solution, and you can get the exact answer in two minutes; in 10 seconds if you are good. Why in heaven's name would you want to use numerical integration to get an approximate answer? So my question is, if that's the exact answer, please tell me what it is to one significant figure. (I'm sure there are people in the room, who, if they thought about it long enough, could get it, but, as a coward, I'm afraid I went to a computer.) The answer to one significant figure is 400,000. If you could have estimated that in your head, I could have made it slightly more complicated, and I think you couldn't have.

Next you'll argue that it is still important that $e^{13}-1$ is exact. You know, for years I accepted that, until one day I woke up, and it hit me that the original integral $\int_0^{13} e^x \, dx$ is also exact! I do not mean that as a joke - I mean that as a deep remark about mathematics, one that I overlooked for a large number of years. Now the question is, if you have two forms that are exact, why is one preferable to the other one? And if you think about that particular example, I think it's very easy to reconstruct the reason. We did not have computers but had tables of e^x. Even today there may be some advantage; you may be able to use a pocket calculator instead of a computer on $e^{13}-1$. But the numerical integration takes less than a second on a computer.

I am not saying that you should not learn how to integrate e^x. But what if it were the integral of x^2e^x? Alright, that's a borderline case, probably still worth doing integration by parts. But if you put in there $x^2 \sin x \ e^x$ I've got absolutely no doubt whatsoever. And it's still less than one second on the computer, and it's the same algorithm that works. You change one line in your program (the formula) and rerun it. Isn't that vastly more important to teach students than horrendously complicated, messy techniques of integration?

Now that doesn't mean I wouldn't teach integration by parts because there are uses in theory and it is part of the beauty of calculus. But I cannot justify about 50% of the content of calculus -- teaching techniques that must be totally useless today. The lesson in this particular case is that mathematicians have got to change their way of thinking from exclusive emphasis on formulas to algorithms, and I heard from Tony that you already had a discussion of this point.

There is only one way, in my opinion, to teach an algorithm and that's to teach it as a computer program. The only way to teach computer programs in a classroom is to have a terminal handy and I intentionally brought up calculus as the example because the case is less obvious there than in discrete mathematics. But to teach the concept of limits, the concept of integrals, the concept of Taylor series and the concepts of differential equations without having algorithms, computer programs and the terminal, seems to me totally irresponsible. Incidentally, Tony, I read the papers you sent me and I basically agree with everything you said. The exception was one small remark you made in passing. You said something like: this conference is not about the use of computers in teaching. I know what you meant by that, but I'd like to convince the audience that while that is certainly a secondary purpose of this conference, the form of the first two years of math teaching and the use of computers in the classroom _cannot_ be separated.

Let me turn back to discrete math, which was my assigned subject. Discrete math is natural for algorithms. It is a natural subject to teach in algorithms and it is an area whose applications are rapidly increasing. Instead of arguing about physical sciences versus social sciences, let me take a different group that happens to be a large part of the clientele at Dartmouth. In freshman math we have a huge number of students who think they are going to be doctors. So what medical schools tell pre-meds that they ought to take has a very big effect on what we teach. And most medical schools insist that you have to take a year of calculus and nothing else. And that's madness.

First of all it's very hard for me to justify that much calculus for the typical doctor. But what really is madness and irresponsible in this complex day and age is to let a future doctor graduate from college without every learning anything about probability theory.

Finite mathematics, however, was constructed in a pre-computer age. It was much later, in the third edition, that our publisher agreed to sneaking in a chapter that mentioned computers, but only if we did it in such a way that any faculty member could avoid using a computer if he or she did not want to use it. They were cowards, and I guess the authors were cowards for agreeing to that condition, because I think that was a terrible mistake.

What I am currently engaged in -- since you asked me about that -- I managed to persuade the department as soon as I got back that it is time to redo the finite math course, as well as redo a programming course aimed at non-computer science, non-physical science students. The finite math and that programming course together are the two courses most heavily elected by Humanities and Social Science students. It was time to redo them by adding significantly more programming to finite math and adding a more significant mathematical content to the other one. Importantly, to make it a two-term sequence. I'm doing it with Tom Kurtz and Laurie Snell, people many of you know.

The interesting thing about the experiment was that the department approved it, as a matter of fact with not terribly much opposition, on one condition -- that we would vouch that teaching materials would exist out of which any member of the department could teach that course! If it sounds funny for very good mathematicians to be worried about teaching what most people in this room would consider awfully elementary material, that reminded me of the mid-50's when finite mathematics came out. It was a fairly revolutionary course: the first review published was very favorable, except it absolutely blasted us for claiming that it's a course that could be taught to freshmen, when it was obviously much too difficult. It had in it, for example, a topic called Markov chains, which, as far as I know, had never been taught in any undergraduate course before it appeared in the finite mathematics book. As a matter of fact, two of the three authors managed to get through one of the best graduate schools in the country without every finding out that there was such a thing as a Markov chain!

So it's obvious that it was a quite revolutionary book, although last week at another Sloan conference, somebody referred to the content of that book as totally trivial and not worthy of the attention of any mathematician. And I took that as a great compliment, because I assure you that twenty-five years ago it was not! We got finite math approved by assuring the department that we would have teaching materials (which they could teach) available. Most members of the department had never in their life had a course in discrete probability theory, to pick one of the topics in it, not to mention mathematical logic of which there is only a baby amount in the book, but they hadn't had that much. With a good book they could teach it; without it they couldn't!

Again, I think there is a very important lesson for what you are doing here. With really good teaching materials easily available mathematicians may be willing to try experiments. Without it they won't, because they don't know enough, and are ashamed to admit in many cases that they don't know enough, to do it.

Finally, Tony asked me to make my comments on the first two-years of mathematics. In a way I already did so by saying that I still agree with what CUP recommended more than a quarter of a century ago. I believe that it makes no sense at all for the vast majority of students to get nothing but calculus. I think if I had my druthers a division with roughly half calculus and half other stuff, including discrete mathematics and learning how to program, would be about right. And I _don't_ want to see it split by the first year being calculus and the second year being the other, because most students never get to the second year. I want both of them split. That may be hard to get. In some way the CUP idea may still be the most saleable: if you could win the battle that the first year is split, basic ideas of calculus and some sort of discrete mathematics/programming course, that would be tremendous progress. And if in the second year there were at least an option for students to go one way or the other, I think it would be a major step forward.

I've got two points of what I would stress if I could attend your whole conference. (But I can't because I just returned to full-time teaching and I taught two classes yesterday and have two tomorrow.) One of the two major things I would argue for is to try to make the case that the calculus that physical scientists really need can be taught in one year. If you can't win that battle, the rest of it won't go. And that you both need better ways of teaching calculus (the computer can help) and that the topics that may have made sense in the pre-computer age don't necessarily make sense today. My second point is the use of computers in teaching. Let me end with a probably poor analogy; but if somebody was asked to teach mathematics in a room which had no blackboard or no vu-graph or no good way of showing formulas in writing, the person would scream and would refuse to teach. I would like to say to you that in this day and age it is equally ridiculous to teach mathematics, particularly certain branches of mathematics, without the availability of a computer terminal. But most mathematicians have not yet realized that fact. Once you get used to teaching that way, it will automatically force you to change your mind as to what is important and how it should be taught.

DISCUSSION

Dubinsky: Do you have any details about what you mean about putting finite mathematics and the first programming course together?

Kemeny: Let me make clear that the result is not intended to be a computer science course. Since finite mathematics naturally leads to algorithms, we are trying to split the emphasis between programming, helping you understand the mathematics, and learning something about programming itself. One of the things we are trying to do is to write a number of computer units that could be used as supplements to our books and others. As examples, with Markov chains you have a chance to teach some nice mature methods and with probability you can do some simulation which is among the more powerful things that students who are not very good programmers can do.

Maurer: What were the reasons why CUP felt so strongly 25 years ago that there ought to be much less calculus?

Kemeny: We became convinced, particularly in talking to social scientists, that there were other subjects in mathematics that were equally important and whose applications were growing very rapidly. Also it bothered me that students in their junior and senior years would see all kinds of absolutely beautiful modern mathematics but that in their first two years students would never meet one single topic that was twentieth century mathematics. Of course, not everything in finite mathematics is twentieth century mathematics but I would argue that finite mathematics represents the spirit of half of what twentieth century mathematics is about.

Pollak: Have you thought about and rejected the notion of intermingling calculus and finite mathematics in the first year beyond just having a semester of each?

Kemeny: I know how to teach a one-semester course in the basic ideas of calculus and its very hard to do it in less. Moreover, I'm not sure what you gain by intermingling. In the sophomore year intermingling seems to make more sense.

PROBLEMS OF IMPLEMENTING A NEW MATHEMATICS CURRICULUM

Stephen J. Garland
Donald L. Kreider
Dartmouth College

Hanover, NH 03755/USA

The fact that this conference is taking place points to ferment regarding the mathematics curriculum for the first two years of college. The present primacy of calculus in that curriculum is being questioned as being appropriate for students in a number of disciplines, especially in computer science. The "twilight of calculus" has been announced in at least one recent interesting and provocative paper [1].

At this conference we have been reminded of the importance of finite combinatorial mathematics, linear algebra, probability, statistics, and computing in the preparation of social scientists, biologists, computer scientists, and others outside the more traditional quantitative disciplines of physics and engineering. We have been called upon to consider the development of a new mathematics curriculum that would better serve computer science and these other disciplines -- a curriculum that would of necessity displace a large portion of calculus from the freshman year of college.

Such proposals are not entirely new. Indeed the development and spread of "Finite Mathematics" in the 1960s was an earlier response to the needs of the social sciences and mathematics. And there has been considerable experimentation with the teaching of statistics at elementary levels, but without much success within mathematics departments.

None of these proposals and efforts has yet produced significant changes in the core mathematics curriculum, i.e., in the sequence of courses designed to prepare students for further study in mathematics. In this paper we explore the reasons why change has been so slow in coming and the problems that must be addressed if a new curriculum is to be implemented.

Background

The main efforts at college mathematics curriculum revision in the 1950s and 1960s, as conducted for example by the Commission on Mathematics [2] and the Committee on the Undergraduate Program in Mathematics (CUPM), were directed toward the needs of students in mathematics and the natural sciences, and they tended to consolidate the traditional role of calculus in the curriculum. Special problems of small colleges in mounting an adequate mathematics program, without the luxury of parallel tracks for different disciplines, were addressed by the General College Mathematics Curriculum (GCMC) Committee of the CUPM [3]. The GCMC recommendations tended to solidify further the role of calculus as the appropriate beginning for college mathematics by making a sequence of calculus courses the common core of mathematics for freshmen and sophomores.

In the early 1970s pressures mounted for alternative beginning mathematics courses, with statistics and finite mathematics being mentioned prominently. The GCMC committee considered this question again and did concede the desirability of alternate entrance courses for mathematics majors and others. But largely because of the perceived lack of "mature" and "smooth" text materials for such courses and, for most small departments, a lack of teachers trained to teach the new courses, GCMC decided that the time was not yet right for a major shift away from the calculus core. It issued only a minor revision of its earlier recommendations.

In part because of this failure to come to grips with the problems of newer disciplines, one of them, statistics, went its own way in the development of non-mathematics based statistics courses for beginning college students. One wonders both whether this was the best solution and also whether it could have been avoided.

New Urgency for Revision

More recently the need for change in the mathematics curriculum has become more urgent in direct response to the needs of the applied sciences, both natural and social, in response to the pervasive impact of computing on mathematics and the sciences, and in response to the needs of computer science itself. Computer science is in fact growing at a rate that threatens to swamp the resources of small colleges and indeed to swamp our national capacity to provide teachers and research faculty in the discipline.

College students in record numbers are demanding access to computer science. College courses in other disciplines are making increasingly sophisticated applications of computing at increasingly elementary levels. And high school students are, or soon will be, prepared in computing to a degree that will make them unwilling to delay a serious study of computer science to the sophomore year. Mere programming courses will not suffice. These students need and want courses that match the intellectual substance of calculus and that apply to computer science mathematical methods using combinatorics, sequences and series, recursion, mathematical induction, and elementary statistics and probability.

The traditional calculus sequence, or even a calculus and linear algebra sequence, is not seen as sufficiently timely or relevant in preparing students for computer science to warrant its being a prerequisite for the first substantive computer science courses. Failure to develop a more relevant and coherent mathematics curriculum for such students can only lead to computer science following the lead of statistics. This is hardly a satisfactory or necessary solution.

Proposals for revision

Various proposals are emerging to revise the beginning mathematics curriculum. They differ in their extremity, in the structure of the proposed curricula, and in the extent to which they target particular groups of students or disciplines. Some such proposals have been made in this conference. These proposals tend to follow one of three approaches:

1. To ignore the traditional calculus sequence (and the disciplines that build their programs upon it) and to develop a completely parallel course or sequence of courses in mathematics for computer science.

2. To develop a single new finite mathematics course aimed at computer science that can be fit into a traditional but curtailed calculus sequence.

3. To modify the existing mathematics curriculum in ways that emphasize methodologies relevant to computer science, but preserving the calculus preparation needed by physics and engineering.

The last approach is a somewhat conservative one and is, in essence, the one recommended in the recent report of the CUPM panel on mathematical sciences. It was aimed at posing few problems of implementation for colleges and students. As a result, it may not go far enough.

Problems of Implementation

Regardless of what proposal for change is made, one is led instantly to a debate which exposes the problems that colleges face in changing to a new mathematics curriculum. Which students will or should such a new curriculum serve? Will a new curriculum be able to serve the clientele of the old curriculum or must schools face the prospect of parallel curricula?

Who will teach the new courses, which may have an expanded computing content -- mathematics or computer science departments? Indeed, who will develop and teach such courses in the absence of any trained computer scientists on the faculty? If more than one department is involved in teaching new courses related to computing, how can course proliferation (such as has occurred in statistics) be avoided? And if parallel curricula are needed and implemented, how do they articulate and at what point do they rejoin? Can small colleges safely or reasonably ignore one of the parallel strands?

Will new curricula lead to the same level of mathematical "maturity" or "sophistication" as the old curricula are presumed to achieve? Will they have a coherence and intellectual depth matching that of the old curriculum?

Will students be left with reasonable choices and career strategies or will they become entrapped by their earlier and possibly faulty decisions? What will be the proper route into the study of mathematics, engineering, or physics? And what should be the presumed common background for upper level courses in these disciplines?

Such questions are real and require solutions. Yet it is our view that debate on these questions is premature so long as the goals, content, and structure of a new curriculum remain vague. It is easy to name a number of topics that should be included in a revised curriculum as well as to name some current topics that could be dropped. But it is hard to reach agreement beyond a superficial level, much less agreement on the competencies that students are expected to acquire through a new curriculum. Without such agreement it is hard to provide convincing answers to any of these questions.

Debate, nevertheless, goes on, taking more or less the form of reiterating the problems implicit in the questions raised above. This debate ultimately focuses on highly specific questions that require decisions at institutional levels such as which students to serve, which departments to provide with resources, which faculty to use, and what prerequisite structure to require. Should recruiting for new faculty be undertaken or should existing faculty be retrained? How to proceed?

Major components in solving problems of implementation

It is clear from previous curriculum revision efforts and from painful experience in many colleges that there are three essential ingredients for success:

1. Texts. It is necessary to have high quality texts that define the courses, that convince faculty of the merits of such courses, and that make it possible to teach the courses with predictable success using diverse faculty.

2. Faculty commitment. It is necessary to have at least one senior faculty member per department committed to getting the courses off the ground. While junior faculty may be more receptive and better prepared for change than senior faculty, they are less in a position to invest the time needed to bring about a major change and less influential in convincing other faculty, both within the mathematics department and without, of the wisdom and long term prospects of the change.

3. Clientele. It is necessary to have a critical mass of students who will elect the new courses, who will use the knowledge they acquire in their major disciplines, and who will be expected to have that

knowledge by the faculty in those disciplines. Without this critical mass, any new course will be short-lived or will remain outside the mainstream of mathematics education, as has happened with previous efforts to introduce such courses (e.g., on finite mathematics).

Without these three ingredients, success is unlikely. Their presence of course does not eliminate the problems enumerated above, but they at least make it possible to approach these problems.

It would appear, therefore, that the proper way to proceed is first to prepare texts for use in the courses of the new curriculum. But what courses? And what authors? What is the basis for agreement on the content and goals for the new curriculum? There appears to be a vicious circle: without quality texts that define the new curriculum, one cannot rationally address the problems of clientele, articulation, and resources; and without agreement on the latter, there is no way to prepare quality texts.

Curriculum development

From the foregoing, it is evident that the problems of implementing a new mathematics curriculum are in large measure the problems of defining what that curriculum ought to be. Hence for most colleges and universities, i.e., for those without visionary textbook writers on their faculty, the problems of implementation are not problems which can be addressed locally, but rather problems which require national attention.

If we can put aside our scruples for a moment it would be useful to examine how corresponding efforts to revise the mathematics curriculum at the primary and secondary school levels have been organized. To a great extent the problems there seem more difficult, if not intractable. Yet progress is being made.

Analogous to the situation in the colleges, there is an entrenched curriculum locked in by existing textbooks and by existing teachers. There are various student audiences to address: college bound, advanced placement, non-college bound, disadvantaged, minorities, learning disabled. There are different vested interests and political factions to recognize and consult: supervisors, teachers, parents and lay groups, school boards, and professional associations. And the problems of small schools in affording multiple track curricula are even greater than those of small colleges.

The vicious circle at the secondary level is broken by separating the curriculum definition process into two parts:

1. Specification of goals in terms of competencies that students or different groups of students should acquire.

2. Development of a curriculum to achieve these goals, including courses, their sequential structure, and text materials.

The specification of goals and competencies requires national impetus, if not consensus. It is here that various interest groups must be intimately involved in the process. Past efforts such as the Cambridge conference [4] and current efforts such as the NCTM *Agenda for Action* [5] and the College Board's *Educational EQuality Project* [6] are examples of this process.

Development of a curriculum that will achieve the specified goals requires, on the other hand, the attention of individual professionals. These will be textbook writers or writing committees who take on the responsibility of implementing specific curricular goals. They will produce course outlines, textbooks, and related materials that develop the content required to achieve these goals together with an appropriate sequencing of topics and experiences that students need to learn.

There is no need for a given curriculum effort to address all possible curricular goals. Indeed schools can and do choose to implement different parts of a comprehensive curriculum, depending on local educational objectives. As long as curriculum materials address a rational subset of the specified goals, they can serve a school choosing to implement just that set of goals. Indeed this is a necessary feature of curricula and curriculum materials, given the diversity of American education.

At the college level there needs to be a major effort paralleling the specification of curricular goals at the primary and secondary levels. This effort must decide what mathematical competencies are needed for computer science, upper level mathematics, and various other sciences or applied disciplines. The interested disciplines must be involved in this process and the competencies must be specified independently of "courses", lest the vicious circle come creeping back.

In the case of calculus, for example, physicists and engineers have long been involved in determining the shape of the mathematics curriculum. Over the years, an understanding has emerged as to which topics are covered at which points in calculus courses and which are taught as needed in physics and engineering courses. Though differences of opinions do exist on the importance, treatment, and placement of various topics, what is crucial is that physicists, engineers, and mathematicians are all committed to and involved with the subject matter of the calculus.

In the case of a new mathematics curriculum, there must be a correspondingly supportive relationship between mathematics and one or more other disciplines. The most likely candidate for one of these disciplines is computer science. What is needed is an understanding of what mathematical concepts are needed for the computer science curriculum, which should be taught in mathematics courses, and which should be taught as needed in computer science courses. Ideally, the relationship between mathematics and computer science should be symmetric; thus there should also be an understanding of what algorithmic concepts are needed for the mathematics curriculum, which should be taught in computer science courses, and which should be taught as needed in mathematics courses.

Clearly such a specification of curricular goals must occur before one can decide, for example, whether all the mathematical needs of students at a given college can be met by a single integrated course sequence or whether parallel courses are required. Given a specification of curricular goals, it is likely that various successful curriculum models will emerge, based on the special needs of certain institutions, and produced by a skilled text writer or group of writers in some of those institutions. What is most important for individual institutions, as they address the question of how to implement a new mathematics curriculum, is that it be possible to compare well-defined curriculum models with respect to the goals they choose to achieve and the particular problems of logistics they engender at that institution.

Recommendations

Given that there is some urgency for revising the mathematics curriculum for the first two years of college, it is undesirable to leave the outcome to chance. Hence it is advisable to undertake the following actions.

1. Conferences or special committees should define the topics that a new mathematics curriculum addressed to particular groups of students must cover. In particular, conferences involving mathematicians and computer scientists should develop an understanding about the interactions between their respective curricula.

2. One or several model curriculum projects at different colleges or universities should write text materials that implement the resulting curricular goals in ways appropriate to the given institution.

Support for the first of these should be sought from government or foundation sources. Support for the second can come appropriately from institutions themselves and from publishers, assisted perhaps by foundation grants. With several model curricula supported by text material developed in this manner, individual mathematics departments would be in a much better position to address the local problems inherent in implementing a new curriculum.

References

[1] Anthony Ralston, "Computer science, mathematics, and the undergraduate curricula in both", *American Mathematical Monthly*, Vol. 88, No. 7, 1981, pp. 412-485.

[2] *Report of the Commission on Mathematics*, College Entrance Examination Board, 1959.

[3] *A General Curriculum in Mathematics for Colleges*, Committee on the Undergraduate Program in Mathematics, Mathematical Association of America, 1965.

[4] *Goals for School Mathematics*, a Report of the Cambridge Conference on School Mathematics, Educational Services, Inc. (Houghton-Mifflin), 1963.

[5] Agenda for Action: Recommendations for School Mathematics in the 1980s, National Council of Teachers of Mathematics, 1980.

[6] Information on Educational EQuality Project is available from Adrienne Bailey, Vice President for Academic Affairs, The College Board, 888 Seventh Avenue, New York, NY 10106.

DISCUSSION

Kreider: Any proposal for a major change in the mathematics curriculum must be accompanied by a detailed model that faculty can examine and judge. It is impossible to visualize what it would be like to teach new courses without a model that included course descriptions and actual textbooks.

We have suggested a two-stage procedure. In the first stage, national discussions would attempt to reach a consensus concerning the mathematical needs of various disciplines; it's not enough just to ask individual engineers, physicists, or computer scientists what their students need. Then, in the second stage, individuals or groups would sit down and develop courses meeting those needs.

It might be possible to circumvent the first of these stages by, for example, having one of the workshops at this conference define carefully a particular model such as that proposed by Fred Roberts in his paper. His model is more feasible than most since it is essentially conservative: it attempts to retain much of the current curriculum while inserting some discrete mathematics. Another conservative approach, suggested by Alan Tucker in his paper, would be to devise new courses for particular clienteles such as computer scientists.

A more difficult approach would be to imagine that the current curriculum does not exist and to try to develop an appropriate curriculum from scratch. Clearly some parts of calculus would reemerge in the first course and later. For this approach to work, the success of the first stage conducted at the national level would be critical. But, as John Kemeny suggested in his talk, this approach might give us the course that pre-med students really need.

Garland: I would like to tell an anecdote illustrating how easy it is to get stuck in the past or in an image of what the past was like. When John Kemeny recently proposed combining two of Dartmouth's courses on finite mathematics and computing into the two-course sequence he described in his talk, many of the standard obstacles to curricular change surfaced: what traditional mathematics would computing displace? could ordinary mathematics faculty still teach the new courses? what would they use as a text? When I suggested that our finite mathematics course had changed little in the 25 years since it was introduced, and that perhaps it was time to rethink that course, one person responded by saying that there had been no need to change calculus courses in those 25 years, and hence that there is no need to change finite mathematics either. Just one of the ironies in this response is that the person who made it was instrumental in introducing a greater emphasis on numerical algorithms into our second term of calculus only two years ago.

A major problem often overlooked in implementing a new curriculum is how to make it stick. Why weren't the hopes of CUP and John Kemeny that finite mathematics would become an integral part of the mathematical mainstream fulfilled? Calculus proved very hard to displace, not just because of its intellectual coherence and longevity, but also because of its built-in clientele. Finite mathematics lacked such a clientele, possibly because it was invented by mathematicians and not demanded by social scientists. Some computer scientists are enthusiastic about including more discrete mathematics in the curriculum, but the computer science community as a whole places low emphasis on mathematics. We may need a national conference, as Don Kreider has suggested, to bring mathematicians and computer scientists together to agree on what's important.

One particular issue they might discuss is how much of the mathematics related to computing should be taught in computer science courses and how much in mathematics courses. Should algorithmic analysis be taught in a mathematics course or in a computer science course? Should sets, relations, and graphs be defined in a mathematics course, with their applications coming later in a computer science course? Having to teach basic mathematics in a computer science course can be a distraction, but a lot of mathematics can be motivated by topics in computer science. For example, it's natural and easy to introduce the O-notation for the asymptotic growth of functions when analyzing searching and sorting algorithms. On the other hand, topics such as proofs by induction, definitions by recursion, and discrete probability might best be taught by mathematicians before students need to apply them in computer science courses.

A final point to remember as we discuss curricular changes aimed at computer science is that we're only a year away from having an Advanced Placement computer science course in high schools. Moreover, many of us involved with this course believe that it won't be long before its subject matter moves down to the ninth or tenth grade level. When that happens, the problem of what mathematics is needed to support computer science will become a problem for secondary schools, not colleges.

Dubinsky: It seems to me that the best way to teach induction and inductive proofs is to have students think in terms of programs.

Garland: I agree, but mathematicians can teach them in terms of programs too. My point is not that we should disentangle what is mathematics from what is computer science; in fact, it's unfortunate that the two disciplines have grown so far apart over the last twenty years. Rather my point is that departments of mathematics will profit from teaching mathematics that is important to computer scientists, and that a good measure of what is important to them is what they are willing to teach themselves.

Maurer: I'm a little disturbed by the implication that the way you should decide what mathematics should be taught by whom is by how long it will take. If it's a long diversion, you seem to be saying, let the mathematician do it. A much better criterion, it seems to me, is whether a piece of mathematics is useful only in computer science or whether it is useful in a lot of other areas as well. In the first year of mathematics we should be trying to attract students to scientific topics generally, and so this is where we should get across ideas that are going to be used in all different fields.

Garland: The question is not how long a topic will take, but how thoroughly it should be taught. Physicists want just enough calculus for their purposes, and that early enough in the curriculum. Likewise, computer scientists want just enough induction and recursion early enough to be of use. Mathematicians, however, would like to develop these subjects more fully, knowing how important they are and how little time physicists and computer scientists are likely to spend on them.

Pollak: You stressed the basic need for textbooks and this is very appropriate. But in meetings with computer scientists, engineers, social scientists, or whatever, what will you have to show? An outline or just some ideas, or what?

Kreider: One would start with a list of the mathematical tools that are used in a discipline and, therefore, of the mathematical topics that might be taught. The purpose of the discussions is to find out which of these topics are essential and which are not. A bit of bartering might go on. Such a discussion would be a great assistance to someone designing a course or writing a textbook. But we should not mislead ourselves into thinking that it would be easy to reach agreement.

PROBLEMS IN INSTITUTING A NEW MATHEMATICS CURRICULUM AT A UNIVERSITY

Lida K. Barrett
Northern Illinois University
DeKalb, Illinois 60115

The role of mathematics in the curriculum

Mathematics plays a central role in the curriculum of any college or university. As a discipline it is not only of intrinsic value and, therefore, worthy of study as a major field, but it is also a generally accepted component of a liberal education, and is usually a portion of or at least an option in, the core or general education requirements for all degrees within an institution. In addition, the subject matter of mathematics--skills, techniques, ideas--provides essential building blocks for the study of an ever-widening set of other disciplines.

The central role of mathematics in the curriculum is unique. English is the only other subject that can claim a universal place in all education. The role of English differs from that of mathematics in that its use in other areas of the curriculum is less dependent on specifics, such as changing techniques, skills, and content. For a discipline other than itself, English is the language for expression of the concepts of the other discipline, while in the case of mathematics, the mathematical concepts themselves often form an essential part of a concept, procedure, or theory of the other discipline. In addition to the ability to use appropriately the fundamentals of mathematics, workers in other disciplines often need to use the content of junior, senior, or graduate-level mathematics courses. Advanced mathematics courses of necessity build on beginning courses, and care must be taken to place concepts to be used in other disciplines within a curricular sequence brief enough to serve the needs of the other area without adding to the curriculum a large amount of unneeded, albeit interesting, material. Even if a mathematics department makes available to other disciplines appropriate mathematics courses, the effect of these courses on the quality and the currency of the remainder of the quantitative portions of the curriculum is dependent on the mathematics faculty building the appropriate interfaces with other departments and programs. Mathematical knowledge and current techniques must become known and adopted in courses outside mathematics, and changes in curricula outside mathematics should have impact on the mathematics curriculum.

The faculty's role in curricular change

The mathematical needs of other areas have been discussed in some detail in earlier papers. The perception of such needs held by the faculty of these disciplines depends to a large extent on their own mathematical training and sophistication. For example, in the biological sciences students trained today are likely to have had a sound calculus course and an introduction to computer science--and therefore these students, when they become faculty members, have the potential to use those tools in their course development, teaching and research. For those trained thirty years ago this is not the case.

The demands that other disciplines will make on the mathematics curriculum will change not only with the increasing mathematical sophistication of their faculties, but also with changing external circumstances. Stephen White, in his essay in The New Liberal Arts [1], states:

> When the culture itself changed slowly, and when liberal arts education itself was a modest enterprise, evolution could take place almost imperceptibly and almost painlessly. To put the matter with perhaps indecent brutality, professors then became extinct at about the same pace as the courses they taught. Today, change, if it takes place at all, takes place a good deal more violently. And there is also a present singularity that is highly consequential: for the time being higher education faces a period of shrinkage for a wide variety of reasons, demography only one among them. Evolution during a period of attrition is never painlessly effected, and cannot be casually undertaken.

The culture is no longer changing slowly, and changes are taking place in many different settings. At one time universities perceived that most changes in the sciences and engineering began in the universities and spread out into the nonacademic world. It is now eminently clear that changes begin in a variety of locations. The cost of equipment, the increasing number of research and development settings outside the academic world, and rapid technological changes have all led to the necessity of curricular changes that reflect not only academic research but also new ideas and even disciplines that have grown to maturity outside of the academic setting. The presence of an informed faculty within a given discipline is essential before curricular change can take place. However, more is needed than this. Donald Kreider [2] gives an example:

> It is difficult even now to predict what would have happened at Dartmouth if it had been proposed as a general rule that every student must take computing and mathematics. On the other hand it is easy to predict the chaos that would have resulted if the faculty had been asked to agree that these are a necessary part of any liberal arts program. Change came at Dartmouth not because new general principles were agreed upon but because the mathematics department, led at that time by John Kemeny, developed a time-shared computer system and a number of specific mathematics courses that students found relevant. The students, given the opportunity, moved to embrace new quantitative skills. The faculty followed in due course.

Two examples from the past

In the recent past, a number of attempted revisions in the mathematics curriculum have had to be abandoned, because faculty members within or outside the discipline have not made the necessary adjustments. In 1968 George Thomas's highly successful *Calculus* appeared in its fourth edition. In this edition, linear algebra was introduced prior to three-dimensional calculus and was used in the presentation of that area. The University of Tennessee, along with a large number of other schools, adopted this text, and rearranged its sophomore calculus curriculum in accordance with the presentation of the book. In courses outside mathematics, however, students were still required to use the more traditional approaches to three-dimensional calculus. The textbook was dropped by Tennessee and by a number of other institutions. In fact, it lost a sufficiently large portion of its market that it was replaced in 1972 in the publisher's offerings not by a revision, but by a revision of Thomas's third edition, identified as the "alternate third edition."

Under the auspices of CRICISAM (Center for Research in College Instruction of Science and Mathematics), in the 1970s a calculus text that included numerical methods and computing was developed. This text, written by Warren Steinberg and Robert J. Walker, was the outgrowth of an invitational conference held in 1970 in Tallahassee, Florida, under the sponsorship of CRICISAM and the Florida State University computing center. A large number of schools adopted the CRICISAM text for some or all of their calculus courses. Again the innovation was not successful. Reasons offered for the failure include the difficulty of finding faculty to effectively teach the material; the reluctance of some primary users (for example, engineering faculty) to add an extra hour of credit or alternatively to drop topics from the standard course; and territorial objections from computer science departments.

Implementing curricular change within mathematics departments

Implementing a new mathematics curriculum on a broad scale, therefore, is dependent not only on the determination of appropriate content but on its relevance to faculty in a range of disciplines and to students. The initial stages in accomplishing such a change are the development of a well-conceived basic course sequence and the addition of this sequence to the curriculum in key colleges and universities. The adoption of these courses will lead to a collection of students and faculty with a common experience, and the process of overall curriculum change within the discipline will have begun. Actually, a significant change in the mathematics curriculum has already begun in a variety of ways in a variety of institutions.

Two papers in these proceedings report on innovative course sequences for the first two years of mathematics. The material in these courses is not new to most mathematicians. Even if it is new to some mathematicians, it is easily accessible to them on the basis of their education and training. Courses developed for the first two years that would replace a portion of the current content of calculus with topics from finite mathematics could, therefore, be taught not only by mathematicians specifically trained in the new expertise but by almost all mathematics faculty. There will, however, be resistance to change because faculty can be expected to display a strong dedication to the value of the calculus, a concern for topics left out, and a lack of familiarity with the new content. However, the case of computer science offers a precedent: almost universally the introduction of computer science courses and their teaching has been carried out in part by mathematics faculty (often self-trained in computer science.)

It appears that a conceptually sound sequence of courses, even one involving a substantial departure from the past in content and in format, could therefore be implemented by existing mathematics faculty. For a large university, if the change is to take place for all students in a given calculus sequence (e.g., all engineers, or all biological science students, or all business students) a textbook, with an instructor's guide and an answer book for the exercises, would be a key to general effectiveness. Further, one or more individuals would have to have the interest and the expertise to serve either formally or informally as resource persons. These two factors would make it possible for faculty and graduate students instructing in the sequence to learn the material as it is being taught or with minimal prior preparation.

The existence of suitable textbook materials is essential for the adoption of a new course in a large department. Teaching from mimeographed notes, from an experimental version of a book, from a book without an instructor's guide and an answer manual might be effective in a small experiment. In an institution with large enrollment and multiple sections, individuals with a variety of past teaching experiences (teaching assistants, senior faculty, part-time instructors, etc.) need unifying textbook material. Such material is necessary in order to be sure that the same content is covered in all sections, and in as consistent a manner as is deemed appropriate.

After a two-year course sequence is in place, the vast majority of the mathematics faculty will be familiar with the content, mode of instruction, and performance of students in the courses. For beginning courses such matters are generally part of the venue of the total departmental faculty, and almost always a subject of regular discussion. For that reason it will be relatively simple for the mathematics curriculum as a whole to be adjusted to take account of the new courses. Upper-level mathematics courses can and will undergo change as the innovations take place at the elementary level.

The exact effect of the changes in the first two years on the overall mathematics curriculum of different institutions will vary widely, just as current curricula vary widely. The <u>Recommendations for a General Mathematical Sciences Program</u> [3] authored by the Committee on the Undergraduate Program in Mathematics addresses in a carefully thought-out and well-documented manner the state of the mathematics curriculum. In Chapter I, "Principles for a Mathematical Sciences Curriculum," the recommendation states:

> The following principles should guide the design of a mathematical sciences curriculum. <u>These principles should serve as a starting point for local discussions to develop a program appropriate to the interests of a particular college's faculty and students.</u>

Further, Chapter II on core mathematics raises the question:

> <u>Question I: Is there a minimal set of core mathematics (algebra, analysis, topology, geometry,) that every mathematical sciences major should study?</u>

The answer given is, in part, as follows:

> Answer: No. There is no longer a common body of pure mathematical information that every student should know. Rather, a department's program must be tailored according to its perception of its role and the needs of its students.

The author, on the basis of her experience at two large universities (25,000+ students), is well aware that the variety of disciplines and the collegiate structure of such institutions significantly affect mathematics department curriculum structure. Northern Illinois University does not have an engineering program, has one of the ten largest undergraduate business programs in the country, and has a large and widely recognized College of Education. These factors have a significant effect on the mathematics and computer science curricula. The University of Tennessee has an engineering school, within which there is a large co-op program, a wide spectrum of doctoral programs in the sciences (including one in mathematics), and so on. Obviously, course offerings at Tennessee differ from those at Northern and will continue to differ in the future.

Implementing curricular changes in other disciplines

Although within the mathematics department most of the faculty will be aware of the content and the presentation of new material, outside the department this will not be the case. It will be incumbent upon the department to communicate in considerable detail to disciplines outside mathematics just what coverage of topics will take place in freshman and sophomore courses, what kind of expertise and skills students can be expected to possess when they complete a couse, and what material formerly covered is no longer covered. It is not enough for mathematics faculty to teach new material unless they present it to the students in a meaningful way and inform the faculty in other departments. Some of the new content of proposed freshman/sophomore mathematics sequences is currently being taught in such areas as computer science, engineering, and physics because it has not been offered in elementary mathematics courses. It is unreasonable to expect other fields to immediately make appropriate changes in their course content, either deletions or additions, or even to expect them to require new mathematics courses that might be suitable for their majors. The span of usage of the CRICISAM and Thomas texts and the Dartmouth experience are good cases in point on this matter. If new material is relevant in other fields, curriculum modification will eventually take place.

The current economic status of higher education, as well as the projected decline in enrollments, means there will be little faculty turnover. In the past, new content has been added to the curriculum by new faculty fresh from graduate school, well versed in the material. Without the regular addition of new faculty a means must be provided for the existing faculty, whose own training has been in the traditional mathematics curriculum rather than the innovative curriculum, to be sufficiently versed in the new material that they are willing to incorporate it in their courses. It cannot be stated too strongly that the mathematics faculty must make an outreach effort to inform the faculty of other disciplines of the exact content covered and the expected levels of students' skill development and retention of material. Professors who joined the faculty in the 1960s and the 1970s, when internal disciplinary focus in all departments (not just mathematics) was so strong that there was little if any interdisciplinary focus or extensive attention paid to service courses, may find this role strange and may expect their overtures to other departments to be received in less than an hospitable manner. Strategies and settings must be developed for implementing external to the mathematics discipline itself, the changing nature of the mathematics curriculum; they will not present themselves naturally.

One strategy for effectively reaching out to other departments will be a thorough and careful study of the use of mathematics in the overall university curriculum. Many universities have moved or are moving toward having either their catalog or their curriculum as part of a computerized data base. Useful information for such a data base includes the number of places a given mathematics course appears as a required part of a major (minor), as an elective part of a major (minor), and as a course prerequisite or corequisite; the number of students in a given major; and the place of mathematics courses in the sequence of the major. Once a study of the use of mathematics in the curriculum has been made, mathematics faculty should contact appropriate faculty in other programs. They should obtain information not only on the specific courses noted but also on the extent to which the mathematics required is used in subsequent courses within the given major. Detailed information and a complete follow-up with the other departments involved are necessary before a change is instituted, in order to avoid making piecemeal adjustments afterward. Further, faculty in other programs must be provided information understandable to them on the content of mathematics courses in order to use this material effectively in their own disciplines. In some cases this may mean that the mathematics department must arrange a series of discussions or short courses for faculty from other disciplines. In other cases the mathematics department may find that the material being added to the first two years of the mathematics curriculum is already being taught in courses in other disciplines. Meaningful interaction with other disciplines, though time consuming, can lead not only to more effective use of mathematics in those disciplines but also to a better understanding on the part of the mathematicians of the uses of mathematics throughout the curriculum.

The curricular process

In most universities the curricular process begins within the faculty, or faculty curriculum committee, of a department. Course changes then progress to the college curriculum committee and thence to a university-wide committee. Most often there are two university committees, one for undergraduate programs and one for graduate programs. Department, college and university committees are often highly political in nature, involving a vote exchange system (you vote for my course, I will vote for yours). The extent to which political activity precludes an honest in-depth discussion of curricular issues is dependent in part on committee leadership and in part on the membership of the committee. Curriculum committees can play a valuable role in helping to implement successful curricular change if they look at the real issues involved, report them accurately to the broad university community in their announcements or minutes, and make certain that effective communication among disciplines and colleges has taken place. It behooves a mathematics department to do its homework not only through effective contact with other departments prior to new course proposals, but within the curricular process itself. Departments should see that curricular changes take place in a statesmanlike fashion. Changes should be accompanied by appropriate documentation of contact made outside the department, and by a clear rationale. Often changes are held up in a committee while its members seek information that is outside their own expertise.

Other issues

From the student's viewpoint, a careful articulation of innovations within the mathematics program and with outside disciplines will mean that change will present few difficulties--that is, if the student enrolls and progresses through the curriculum in a regular manner. However, for the student who stops out, or for the transfer student, careful attention to on-campus curricular matters will not be enough.

Universities that regularly receive a large number of community college transfers will need to inform nearby community colleges of impending curricular changes, and to work with them so that, if at all possible, these schools can make changes in their course offerings. It should be noted, however, that most community colleges serve as feeder institutions for more than one senior college, and care should be taken to provide a variety of options and accommodate different ambitions, otherwise; community college transfers may find transfer unsatisfactory.

Anyone who has previously dealt with broad curricular changes, or with changes in course outlines or textbooks, could add to the above list of concerns. Among other issues are the necessity for trailer sections as the curriculum is phased out; the problem created for a student who fails a semester of the curriculum; and the effect of co-op programs (in which students work alternate quarters or semesters) on the implementation of changes. All these factors--and others--contribute to the inertia against change.

An issue different in nature from the above, one that cannot be ignored, is the faculty time commitment needed to bring about broad curricular change. The development of a new course is in itself time consuming. The mechanics of implementation of the course within the department adds an additional time burden. The effective implementation of the course so that it becomes a meaningful part of the curriculum of other departments requires a further and large-scale time commitment. The time invested by faculty members working with other disciplines needs to be specifically noted and encouraged. The department and the college in which the department is located must be prepared to provide time for all these activities and, when the changes take place in a satisfactory manner, to provide rewards in terms of recognition and respect for the individual faculty member involved and for the department.

An historical perspective

Change in the content of the curriculum, of course, is an ongoing process. It is played out again and again within the undergraduate course of study. It is perhaps helpful to go back and look at the very earliest days of the mathematics curriculum in this country [4]:

> Mathematics was only indifferently included in the eighteenth-century Harvard curriculum, a slight remnant of the ancient quadrivium, considered perhaps of use to mechanics but of no value to gentlemen, scholars, and men of affairs. A little arithmetic and plane and spherical geometry, the latter read in English as if even further to reduce its importance, appeared in the senior year almost as an afterthought.
>
> Mathematics experienced a significant change of fortune in the curriculum in 1714 as the result of a collection of over seven hundred books gathered in London by Jeremiah Dummer. . . .Dummer had solicited books from Isaac Newton and other leading figures of the Royal Society, an inspired moment in the history of philanthropy that led Yale's first two tutors, Samuel Johnson and Daniel Browne, both of the class of 1714, to insinuate Copernicus, Descartes, and Newton into the curriculum, so carried away were they by what they found themselves reading in the Yale library. Their lectures and conversations with their students soon made clear that the "New Learning," which had arrived in Dummer's parcels of books, could not be understood without more mathematics than the meager arithmetic with which students entered Yale. In 1718 algebra appeared in the Yale course of study, in 1720 astronomy was being studied in mathematical terms. . . .In 1742 mathematics instruction began in the freshman year; three years later geometry appeared as a sophomore subject and mathematics was included in the junior year as well. The growth of mathematical studies in the curriculum was unavoidable once Newtonian physics made its way into the course of natural philosophy.

Obviously, significant changes in our culture have had and will continue to have an impact on the curriculum of the mathematical sciences. Change cannot be avoided. A strong methematics department will be aware of appropriate changes that should be studied and implemented not only within its own curriculum but throughout the university's undergraduate curriculum. An awareness of the magnitude of the problem of change can lead not only to its solution but also to fruitful interaction with other disciplines.

References

1. Stephen White, "The New Liberal Arts," in The New Liberal Arts: An Exchange of Views, ed. James D. Koerner (New York: Alfred P. Sloan Foundation, 1981), p. 3.

2. Donald L. Kreider, "Response 8," in The New Liberal Arts, p. 60.

3. Committee on the Undergraduate Program in Mathematics, Recommendations for a General Mathematical Sciences Program (Washington, D.C.: Mathematical Association of America, 1981), pp. 12 and 44.

4. Frederick Rudolf, Curriculum: A History of the American Undergraduate Course of Study since 1636 (San Francisco: Jossey-Bass Publishers, 1977), p. 33.

DISCUSSION

Barrett: From my position as Associate Provost at Northern Illinois University I can see that what we are doing here in mathematics is part of a larger picture because similar things are happening in other majors and, of course, in the undergraduate curriculum broadly. Four sources of interesting and useful data on this problem and its history are:

The New Liberal Arts, Sloan Foundation, 1981

Frederick Rudolph, Curriculum: The History of American Higher Education Since 1726, Josey-Bass, 1977

Arthur Levine, Handbook on the Undergraduate Curriculum, Josey-Bass, 1979

Quest for Common Learning, Carnegie Foundation, 1981 (Videotape and book)

At a large university like mine the faculty is too big and there are too many different programs to institute a new mathematics curriculum by what I would call "everybody knowing everybody" techniques. As noted in the previous paper, you're going to have to have texts. And you probably need a teacher's guide and an answer book for the homework. And in a large department you need some key people who favor the change because one problem you will always have is that everybody in the mathematics department is going to fight over what is being taught in the freshman and sophomore courses.

We also have the problem of faculty readiness in other disciplines. These faculty will be held back by what mathematics they were taught. One possibility here is to teach appropriate mathematics courses just for faculty - not students - in other disciplines. We did this once successfully with biologists at the University of Tennessee. We either have to retrain faculty or wait another generation for real change.

And we must pay attention to the politics that goes on with curriculum. You must honor the curricular process, whatever it is, at your university. We have built walls between departments, we have built the territorial imperative, we all think that money follows credit hours and our structures are not set up to do the kind of dialogue and contact with other programs that we need right now. On the other hand, lack of faculty mobility, usually thought of negatively, can be a plus here because, whatever changes we make, we will be dealing with the same people for some years. Also the realization that there will be little or no growth for some years could lead to better utilization of resources and even perhaps to more cooperation.

Finally, I'd like to reemphasize a point made by John Kemeny that, whatever new courses we may start, they will die if we don't build bridges to other disciplines.

Ralston: Do you agree that, while a small college might develop a text and a new curriculum and almost immediately implement it in a pilot program, this is just not on at large universities like Northern Illinois?

Barrett: You could do experimental courses and develop textual materials in these courses. But for a large university, I don't see that you could handle a general change in the mathematics curriculum quickly.

ARITHMETIC IN THE COMPUTER/CALCULATOR AGE

R.D. Anderson
Louisiana State University
Baton Rouge, LA 70803

This conference focuses on the possible or likely effects the computer/calculator age and the discipline of computer science will have or should have on the first two years of the college and university level mathematics curriculum. Related and perhaps even more fundamental effects will be those on pre-college mathematics. Any such effects will generate further effects on beginning college level mathematics including, of course, remedial mathematics.

It is the purpose of this paper to discuss some of the likely effects on arithmetic of the pervasiveness of the computer, the calculator and technology in our society. There will also be effects on various high school subjects but the effects on arithmetic and more generally on quantitative thinking and processes will be earlier and perhaps more important.

Another major issue of the school curriculum will be the need to expose students to computers and calculators and more generally to modern technology, to acquaint them with their uses and roles in society, and more generally to equip all students to live in the quantitative and technological age now developing. A further major issue is the developing use of computer and other technology in the educational process, a trend which may be greatly accelerated by the apparent inability of our society to develop and maintain an adequate supply of competent mathematics and science teachers.

But back to the issues of arithmetic in the computer/calculator age.

<u>A bit of historical perspective</u>:

The nature, scope and sequencing of student exposure to arithmetic in this country was established by the publication in 1834 of Ray's textbook, comparable to McGuffey's readers in importance to its subject area. In many respects arithmetic is unchanged to this day. It became one of the three "R's" in grades one to six. When grades seven and eight were added to the curriculum, further arithmetic was taught including applications to practical problems and review of earlier learned arithmetic.

When grades 9-12 became common, algebra, geometry and trigonometry were taught with little arithmetic in those grades. Gradually up to two years of applied arithmetic (for example, ninth grade general math and twelfth grade senior arithmetic) were offered for students not ready or able to handle the usual more abstract college preparatory mathematics. Now a growing number of colleges and universities give placement tests and feel forced to offer remedial arithmetic--e.g. in the fall of 1981, Louisiana State University started offering remedial arithmetic with enrollments of about 40% of the total number in the freshman class.

<u>Factors inhibiting change</u>:

The content and emphases of school arithmetic are determined largely by the textbook series actually used in classrooms and by teachers' selections from material and exercises in these textbooks. Textbook series adoption procedures vary from state to state, district to district, and school to school but judgments by supervisor and teacher committees are almost certainly the primary determinant of which texts and what type of texts get used.

Textbook series strategies are designed by publishers, editors, groups of authors and consultants with the primary purpose of producing texts that will sell, i.e. be put on approved lists and be used in the classroom. Obviously, all concerned like to believe that their series have good educational value but innovations are more designed for sales than for student learning per se. Another major factor affecting the selection and emphasis of material taught in the classroom is the growing use of standardized tests, interpreted variously for individual, class (and their teacher), district and state evaluations. The press regards test results as important educational news. The nature and emphases of problems in these tests tend to be a conservative force in curriculum revision but conceivably could be used to encourage new emphases if test producers can be encouraged to modify their tests in accordance with changing societal needs.

Another factor discouraging changes in emphases in arithmetic is the "motherhood" attitude concerning what is important. What was good for the parents will be good for the children; "back to basics"; the verities of the three R's are fundamental and transcend time.

The import of the preceding comments about factors that inhibit change in education are not meant to discourage us or force us to give up before we start--rather they are intended to identify the points in the system which must be faced if we are to encourage useful change.

The changing nature of arithmetic

What are student needs in arithmetic for further learning and functioning in the computer/calculator age? What aspects of arithmetic will be more important, what less so?

1. First and foremost, quantitative understanding and thinking will be more important than ever before. In an increasingly complex technological society, numbers and their uses become more fundamental for more people.

2. Single digit number facts are at least as important as ever. Thorough familiarity with sums and products of single digit numbers is vital as are the implied subtraction and division facts e.g. $14 - 9 = 5$ and $42/6 = 7$. In the past, these facts have been important so they could be applied in algorithmic contexts such as column addition, several place multiplication, and long division. Now

knowledge of the number facts will be needed in broader contexts, particularly those not involving paper and pencil, in informal mental arithmetic, in estimation, in sensing or testing reasonableness of answers displayed on machines. Texts (and teachers) have a tendency to teach facts primarily in special formats applicable to later paper and pencil algorithms. Indeed according to Steve Willoughby, President of the National Council of Teachers of Mathematics, a study has shown that 93% of students' learning of arithmetic is with paper and pencil and 90% of peoples' uses of arithmetic already does not involve paper and pencil. In the future there will be almost no paper and pencil arithmetic (except for exercises in classes). We will be forced to teach people to think arithmetically.

3. The use of paper and pencil algorithms will become almost obsolete. However the concepts of "algorithm" (and of various several digit arithmetic algorithms specifically) become more important while drill work on the use of paper and pencil arithmetic algorithms becomes largely irrelevant in terms of future student real needs (as distinct from transitory test taking needs). These numerical algorithms were developed over the centuries following the advent of the decimal system. They vary by culture, are highly refined and are quite efficient for the purpose of making detailed and accurate calculations. They are learned by students as processes--analagous to formulas. Most users either do not understand them or use them automatically with no thought of understanding. They are a substitute for thought, and unfortunately many people and even some teachers regard their use as an (or the) end goal of arithmetic. They do have one very useful virtue and role--they reinforce the single digit number facts by constant use of them (but in a specialized format). As the paper and pencil algorithms are deemphasized in school, we shall have to provide alternate devices for practicing and review of the basic facts. We should also seek better student understanding of the principles of the algorithms. But we can sacrifice some efficiency for better intuition. For example, without calculators bracketing square (or other) roots between successive integers, tenths, hundredths, and so on is much less efficient but much more intuitive than the traditional square root algorithm. With calculators, the bracketing process is efficient and can also be used to teach informal interpolation, at a time when formal interpolation in tables is becoming irrelevant.

4. Algebraic manipulations of arithmetic fractions are becoming much less important. To many students, fractions do not represent numbers, they represent objects on which certain operations or algorithms are performed to produce something called "the answer." Yet in the computer/calculator age, there will rarely be any use for adding, or subtracting, or dividing numerical fractions--there probably will be some use for multiplying fractions. In algebra, operations on fractions will probably still be important but the algorithms for algebra can probably best be taught in algebra classes to the (slightly) selected students, with justifying

explanations from arithmetic. With symbolic manipulations now possible on a computer, there are some who question whether the current emphasis on algebraic manipulation makes sense. I am not prepared to judge this issue, there are too many side effects for confidence in any easy or quick judgment. In any case, students must be taught to think of numerical fractions as representing numbers, not as abstract entities on which one performs operations to produce an answer. Here are four illustrations. In a second semester calculus class, a colleague of mine got an answer of 164/16, whereupon he remarked that the answer was a bit more than 10. Hands shot up, "How do you know?" Apparently 164/16 is not a number to many students, rather it is a symbol on which they perform an operation (division of 164 by 16) to produce an answer. A former colleague laboriously got a college calculus student to produce 105/7 as an answer whereupon it was revealed that the student could not evaluate it or estimate it without a calculator. Twenty five years ago, a football player (not an interior lineman) on a national championship team gave a test answer that $1/2 + 1/3 = 1/5$; thus he even messed up the wrong algorithm. Obviously to him the fractions had no meanings as numbers. On an ETS test administered to about 550 trigonometry level high school students, about 25% missed the question "Given $a = 1/2$ and $b = .49$. Then (i) $a = b$, (ii) $a < b$, (iii) $a > b$, or (iv) there is insufficient information to tell." If many trig level students don't think of fractions in terms of their decimal equivalents, we are in a sorry state. In the age of technology, numerical fractions will automatically be converted to decimal approximated form.

In a standard and widely used seventh and eighth grade textbook series there were a dozen or so exercises on comparing fractions, but not like 2/5 and 1/2 or even 17/35 and 1/2, rather like 7/12 and 19/35 and the method shown was to "cross multiply and compare," another unmotivated and unexplained algorithm.

A (or the) standard method of adding fractions is to find the least common denominator, convert each fraction to a form involving this denominator and then add numerators, simplifying if necessary. Are the LCM and GCD still important as arithmetic concepts for all students to know? We teach the LCM and GCD primarily for their use in efficient adding of numerical fractions. The arithmetic LCM process is of little value in the algebraic adding of fractions.

Is the division of numerical fractions important to teach and emphasize to all students? There is almost no practical application for the dividing of numerical fractions except for conditional probability where it can be taught separately (and easily to select students). In solving equations like $3/5\ x = 4/7$, one can more easily multiply by 5/3 than divide by 3/5. After all the multiplication by 5/3 illustrates the rationale for the "invert and multiply" algorithm.

Without a careful study of time allocations, I would guess that fractions generally would get almost the same time as at present but with radically different emphasis and exercises. Thus students need emphasis on comparing fractions as numbers followed by much informal estimation (e.g. 12/25, 21/39, 13/40, 149/301), adding (or estimating addition of) mixed fractions by whole numbers first and then by proper fractions, conversions to and from decimal equivalents and/or approximations, estimates or bracketing of sums, differences and products, mental processes for handling "easy" fractions, etc. A recent British study has shown that outside of school, few people ever add fractions except in the special cases of denominators which are powers of 2. We need a very hard look to determine those aspects of arithmetic relative to fractions which are really important for practical use, for background for other mathematics and for computer use.

5. Decimals and decimal approximations are becoming much more important. Calculators and computers will spew out decimal approximations to much greater accuracy than is generally needed. Intelligent or common sense rounding off will become more important as will, for better students at least, the distinction between decimal representations and decimal approximations. Conversions from and to decimals as well as order of magnitude estimates are going to be basic. The understanding of 24.837 as $2 \cdot 10 + 4 \cdot 1 + 8 \cdot (.1) + 3 \cdot (.01) + 7 \cdot (.001)$ and as $2 \cdot 10^1 + 4 \cdot 10^0 + 8 \cdot 10^{-1} + 3 \cdot 10^{-2} + 7 \cdot 10^{-3}$ becomes more important as does the use of rounded off estimates to determine if the displayed answer is in the right ballpark. The algorithmic paper and pencil processes involving decimals become much less important. Generally speaking, familiarity with decimals, their uses and their limitations is part of the wave of the future.

6. Percents need to be understood and estimated, not computed by formal algorithms. Surprisingly few students realize that 1½ percent of a number can be computed by taking 1 percent, half of that and adding them together. Rather they have been programmed to multiply the number by .015, a tedious and potentially error producing process. More generally, students are given algorithmic and not intuitive ways of handling percents. At a Holiday Inn recently, with a room charge of $45.00, I asked for a 10% discount on the basis of my AARP(NRTA) card. The nice looking clerk called another clerk over to help her figure out what I should be charged, they punched some figures into a hand calculator and told me the bill was $40.95. I never did find out whether they thought 10% of 45 was 4.05 or whether 45 - 4.50 was 40.95 or whether there was some other explanation. I didn't argue, the story was worth more than 45¢. When young people who are hotel clerks have trouble with deducting 10% of 45 from 45, our educational system for the age of technology is in real trouble.

7. Estimation and approximation of numbers in many forms is very badly needed. Reasonableness or "ball park" criteria for answers are, of course, fundamental to the intelligent use of numbers, particularly when they appear on displays with no rationale for their existence. In the standard textbooks which I have looked at, estimation is done almost totally via formal round-off, (another formal algorithm) and is rarely used in other problems. Indeed, there are two quite different estima-

tion concerns that should be used--one the application of numbers to physical, geometric, fiscal or everyday life applications. It should not take 1200 gallons of paint to paint an ordinary house, it should not take 10 days to drive from New York to Chicago on the interstate, the ratio of the area of a circular region to a circumscribed square region is certainly between ½ and 1 and in fact is pretty close to, but less than, 4/5, the social security system is bound to have problems when soon to be eligible recipients, like myself and family, will get annual payments of more than 1½ times the total I put in over the years.

The other needed aspect of estimation is strictly numerical, the estimation and use of numbers when they arise. In spite of a virtual absence of meaningful precise numbers in ordinary use in our society (except perhaps in bank accounts and bills) we encourage almost no student emphasis on estimation. The issue rarely if ever comes up. Some examples have been discussed previously in this report. Here is another. In possible conversion to the metric system, people need a preliminary feel as to what numbers mean in relation to their experience. They don't need a formula for conversion but rather landmarks. They certainly don't need four place accuracy but convenient estimates. Thus a liter is about 5% more than a quart, a kilometer is about 3/5 or 5/8 or 8/13 of a mile, a kilo is about 2.2 pounds, a temperature of 68°F is equivalent to 20°C while 32°F is equivalent to 0°C, more generally a change of 9°F is equivalent to one of 5°C.

8. Informal mental arithmetic is going to come back into its own. Mental arithmetic has virtually disappeared from texts and from classrooms. We certainly do not need fancy or sophisticated mental arithmetic but we need enough to let us estimate and use numbers intelligently. Thus in adding 37 and 19 we should add 37 and 20 and then subtract 1, which is much easier and less error prone than a mental use of the usual algorithmic process. For 96x237, if we only want to be within 5% of the answer we could cite 100 x 237 = 23,700 or 23,500 or even 23,000 or 22,700 if we want to be more accurate. Except in class, we rarely need to be more accurate than that. Also 22x18=396, $((20 + 2)(20 - 2) = 400 - 4)$. Texas has, for years, had a University Interscholastic League competition on "number sense" or mental arithmetic, literally 80 questions in 10 minutes with no scratch paper or extra writing permitted. The number sense tests are designed at various grade levels. National versions of these competitions might be very healthy in encouraging elementary and middle schools to emphasize numerical awareness.

In mental arithmetic, the issue is to find an easy, not a formal algorithmic way to produce an answer. Thus adding or subtracting across 100 (or a multiple of 100) can frequently easily be done by subtracting from or adding to 100. For example, $87 + 19 = 87 + (13 + 6) = (87 + 13) + 6 = 100 + 6$, obtained by borrowing 13 from 19. It could also be done as $87 + (20 - 1) = (87 + 20) - 1 = 107 - 1$. The important issue is finding and using an easy method. Also factoring may provide easy processes for multiplication. Thus $7 \cdot 14 = 7 \cdot (7 \cdot 2) = (7 \cdot 7) \cdot 2 = 49 \cdot 2 = (50 - 1)2$

= 100 - 2 = 98, and $15 \cdot 12 = 5 \cdot 3 \cdot 2 \cdot 6 = 5 \cdot 2 \cdot 3 \cdot 6 = 10 \cdot 18 = 180$. Adequate exercises in texts will encourage students to try such processes and, in so doing, review and practice more basic number facts.

Another aspect of informal mental arithmetic is that contrary to some prevailing teacher attitudes there is no unique best (or even clear) process to produce an answer, there are almost always many processes which can produce valid answers to any problem. This fact may jeopardize both the authority of some teachers and their reported unwillingness to listen to students' ideas. Yet it is precisely the availability of various approaches and the acceptance of them which opens the door to much healthier student attitudes toward understanding and effective use of mathematics. To achieve both an openness to other methods and a willingness to accept common sense estimations will require a rather long term teacher and public reorientation. But both are fundamental to better student understanding of arithmetic and mathematics.

9. Applications of arithmetic and problem solving will become more important. The computer and/or calculator reduce the need for student time spent on calculation and thus increase the time available for the uses of arithmetic, i.e. problem solving. The computer can and should be used to let children get numerical answers to real world problems and to decide if they are meaningful. But we need an approach to problem solving that goes beyond the formal and frequently memorized approach to setting up the question algebraically and then solving it algebraically using rigidly structured formats.

My suggestion would be to first get students to guess at answers, to use trial and error approaches to questions for which these approaches will (usually) produce answers. Make students first learn to read and understand the problem without either setting it up algebraically or solving the resultant algebraic system. We need to encourage feel for numbers as they apply to worded problems, not formal systems which produce formal solutions. While this topic is principally outside that inherent in the computer/calculator revolution per se, it is not outside the philosophy of encouraging and developing thought as distinct from formal processes in arithmetic.

10. Mensuration and other applications of arithmetic to geometry need rethinking. Along with downplaying the role of memorized computational algorithms we also will need to downplay the role of memorization of formulas and the automatic unthinking use or abuse of formulas. In geometry, for example, we should stress basic formulas like that for the area of a rectangular region and the related one for the area of a triangular region (with many pictorial representations and alternate form uses), the area of a circular region (with intuitive explanations as to why it makes sense numerically), and the process of decomposing more complicated regions into more elementary ones whose areas are already known to be computable. For example, the usual formula for the area of a trapezoidal region is almost totally

useless.

At an earlier stage, lengths of arcs can be produced from the Pythagorean Theorem and the formula for the circumference of a circle. Students need many exercises which force them to decompose more complicated unions of linear segments and circular arcs into their known component parts. The principle of deliberate reduction of complicated problems to simpler (or simple) cases is vital in the mathematical process and appears to be almost totally ignored in most texts.

For volumes of three dimensional solids, we should use only the most basic formulas such as area of the base times height for cylinders or parallelepipeds and the corresponding formula for a cone or pyramid. For surface areas, we just reduce to simpler known (planar) cases, thinking of the geometry involved. Only the better students need be expected to know the surface area or volume of a sphere.

It is important that the geometric intuition leading to area and volume formulas be emphasized as well as miscellaneous counting procedures for various geometric figures. Indeed, counting procedures for geometric figures are good both for geometric intuition and for an introduction to more abstract discrete mathematics.

Concluding Statement:

We should seek to identify processes by which we in the mathematical and scientific community might hope to effect changes in emphases in arithmetic of the sort identified above. My own suggestion is that we start a public dialogue on the issues, stressing that the computer/calculator age does change the nature of those aspects of arithmetic which are important for students to know. Such a dialogue can conceivably lead to changed public, teacher and textbook editor conceptions of what arithmetic should be. I see no realistic hope of effectively improving arithmetic performance without such initiatives. Unlike the new math, change can probably be accomplished by incremental changes specifically exploiting the nature of arithmetic exercises in texts and of problems on standardized tests to help encourage teachers to change emphasis from computational to intuitive arithmetic.

A LOWER-DIVISION MATHEMATICS CURRICULUM CONSISTING OF A YEAR OF CALCULUS AND A YEAR OF DISCRETE MATHEMATICS

Richard D. Alð (Lamar University), D. Bushaw (Washington State University), Donald L. Kreider (Dartmouth College), Jack Lochhead (University of Massachusetts), William L. Scherlis (Carnegie-Mellon University), Albert W. Tucker (Princeton University), and Julian Weissglass (University of California, Santa Barbara)

The authors considered the problem of writing a curriculum for the first two years of college mathematics that would consist of a year of calculus and a year of discrete mathematics. They imposed the further constraint that it should be possible to take either year alone, and therefore to take both in either order. It was recognized that this constraint exacts certain costs, such as some need for repetition and fewer opportunities for cross-references; but it was intended that the additional flexibility should provide for the needs of a wide variety of undergraduate students, with the understanding that in many cases (e.g. for students concentrating in mathematics, computer science, engineering, or certain physical sciences) additional courses would be needed.

The rationale for curtailing traditional calculus to allow the inclusion of a rather broad introduction to discrete mathematics has been presented elsewhere and will not be repeated here. It should be said, however, that the authors would like to see a considerable amount of computer (or calculator) use by students and instructors in all parts of these courses. What forms this activity takes will depend on local circumstances, so will not be made specific here, but should be considered an essential part of our recommendations.

It should also be said that this report is the result of only one day's intensive discussion, and should therefore be treated as even more tentative than it might otherwise be. For the same reason, the course outlines are less detailed than one might wish.

I. DISCRETE MATHEMATICS

Certain general themes should pervade this course, although not all of them will be mentioned explicitly in the outline. Typically, these themes might first be brought in almost casually, then presented more distinctly and emphatically in several interludes where the instructor would set aside one or several consecutive class meetings to review topics recently considered from the standpoint of the general themes.

Themes to be treated in this manner include:

recursion	multiple representations
induction	diagnosis of problems
modeling	the process of abstraction
algorithms	the nature of proof

care in the use of notation

Proofs should be presented only when they are brief and especially instructive; proofs of an algorithmic nature deserve special emphasis.

Among specific topics that might have been considered for inclusion in such a course are the following:

formal languages	linear programming
automata	planarity of graphs
block designs	finite fields
determinants	computational complexity

These have been omitted from the outline because, in the opinions of the authors, they would more properly be covered in a more advanced or more specialized course, or are simply less important than the topics included.

We firmly believe that this course should be taught at about the same intellectual level as the calculus. There is nothing intrinsically elementary about discrete mathematics, and, properly taught, it should do as much as any other kind of mathematics to develop a student's mathematical maturity.

The title of each major section of the outline is followed by a rough indication of the percentage of the year's work that might reasonably be devoted to the material in that section.

OUTLINE

1. Graphs, logic, and sets (20%)
   - graphs
     - fundamental ideas
     - connectedness
     - trees, circuits
     - search algorithms
   - logic
     - propositional calculus
     - quantifiers and their properties
     - Boolean algebra
   - sets
     - fundamental relations and operations
     - functions, mappings, relations

representations, isomorphism

2. <u>Induction and recursion</u> (15%)
   - induction
     - elementary finite induction
     - inductive proofs and definitions
     - recursive definition of sequences
   - recursion
     - general discussion, with further examples
   - recurrence relations
     - differencing and summation
     - closed form solutions of linear homogeneous recurrence relations with constant coefficients
     - applications: Fibonacci sequence, binary search
3. <u>Counting and probability</u> (20%)
   - counting
     - permutations and combinations
     - binomial and multinomial theorems
     - inclusion/exclusion formula
   - probability
     - elementary finite probability
     - Monte Carlo method
4. <u>Special relations and semigroups</u> (10%)
   - order relations
     - partial, quasi-, and linear orders
     - well-founded ordering
     - applications, e.g. in social sciences
   - equivalence relations
     - definition
     - partitions and equivalence classes
     - abstraction by equivalence
   - semigroups
     - associative binary operations
     - identities and monoids; inverses
     - isomorphism of semigroups
     - examples: Cayley tables, composition of functions, strings, set operations
5. <u>Matrices</u> (20%)
   - matrix algebra, including row and column vectors
   - linear systems, Gaussian elimination
   - inverses
   - linear independence and ranks
   - the space of real n-tuples
   - subspaces, bases

inner products, orthogonality

6. Statistics (15%) (This is only an introduction. A separate one-term course in statistics would be preferable.)

descriptive statistics

mean, median, mode, correlation, regression lines

distributions

binomial distribution

normal distribution as approximation to the binomial

confidence intervals and hypothesis testing (if time permits)

II. CALCULUS

To some extent, the themes listed earlier for the Discrete Mathematics course should also be emphasized in the Calculus course. Although at first sight the outline given will look rather conventional, closer examination will show that several traditional topics are omitted, and some topics that usually receive a good deal of attention are markedly deemphasized. What does not show is our intention that intuitive aspects, numerical approaches (with computer or calculator implementation), and illustrative applications should play an important role throughout.

Some topics that might have been expected in this outline but have been omitted are listed below. In each instance, it was our opinion that the topic could not be included in a reasonable one-year calculus course without sacrificing material of much greater importance to most of the students.

rigorous theory of limits

proof of existence of definite integrals of continuous functions

special techniques of integration

such applications of integrals as force on a surface, centroids, moments of inertia

detailed treatment of calculus of inverse circular functions

alternating series; conditional convergence of series

Taylor series of functions of several variables

Lagrange multipliers

vector analysis

special techniques for solving first-order differential equations in closed form

L'Hôpital's rule

hyperbolic functions

The authors believe that a student who has mastered the material outlined below should experience little difficulty in picking up most of the omitted topics as needed.

The six major sections of the outline should be given approximately the same amount of time in a one-year course.

OUTLINE

1. Limits, continuity, derivatives
    - sequences (the concept, convergence)
    - limits (presented intuitively) and continuity
    - definition of derivative; differentiation as a mapping
    - interpretations of the derivative (slope, velocity, marginal quantities in economics, etc.)
    - techniques of differentiation, including composite and inverse functions
    - higher derivatives
2. Applications of the derivative and related ideas
    - maxima and minima
    - curve plotting
    - the theorem of the mean
    - differentials
    - Newton's method
3. Fundamentals of integral calculus
    - approximating areas by summing areas of rectangles
    - definition of the definite integral
    - trapezoidal rule, midpoint rule, Simpson's rule; the relation between them
    - antiderivatives
    - fundamental theorem (two forms)
    - brief account of techniques of integration (polynomials, substitution, parts; use of tables)
    - applications (e.g., areas, volumes, averages, work, cumulative income)
    - definition of the natural logarithm and (as its inverse) the exponential function; their derivatives and integrals
    - examples of improper integrals, with emphasis on evaluation
4. Differential equations
    - $y' = ky$; growth and decay
    - the differential equation $y' = f(x, y)$
        - geometrical approach
        - numerical methods of solution
    - differentiation and integration of sin and cos, with remark that other circular functions may be treated in terms of these
    - the differential equation $y'' + ay' + by = h(x)$
        - general solution
        - use of initial conditions
        - applications

5. Power series; calculus of functions of several variables, I
   - infinite series
     - definition of convergence
     - convergence tests only as needed for discussion of power series
   - power series
     - radius of convergence
     - manipulation, use in approximation
     - Taylor's theorem with remainder
     - generating functions
   - polar coordinates
   - three-dimensional coordinate geometry; rectangular, cylindrical, spherical coordinates
   - multiple and iterated integrals

6. Calculus of functions of several variables, II
   - partial derivatives, definition and interpretation
   - the chain rule
   - total differential
   - critical points of a function of several variables and their relation to extrema
   - directional derivatives and gradients
   - level sets, tangent hyperplanes
   - vector fields through two-dimensional Green's theorem (definition of line integral).

AN INTEGRATED TWO-YEAR CURRICULUM--REPORT OF A WORKSHOP

Lida K. Barrett (Northern Illinois University), Isaac Greber (Case Western Reserve University), Robert Z. Norman (Dartmouth College), H. O. Pollak (Bell Laboratories), Ronald E. Prather (University of Denver), Fred Roberts (Rutgers University), Herbert S. Wilf (University of Pennsylvania), Stanley Zionts,(SUNY at Buffalo)

Introduction

An integrated two-year curriculum in mathematics was developed during a one-day workshop session. Please note that this paper is precisely what its title states--the report of a workshop. The workshop followed two days of discussion of the papers presented earlier in this volume. Drafts of these papers had been read by the participants prior to the conference. Nine papers contain recommendations related to course content for the first two years. These papers are those of Alo, Bushaw, Norman, Roberts, Scherlis and Shaw, Weissglass, Wilf and Zionts. Further, the Recommendations for a General Mathematical Science Program of the Committee on the Undergraduate Program in Mathematics [1], served regularly as a basic resource in the discussions.

The notations adopted during the workshop are used. They were not refined to a mnemonic scheme, nor were other refinements made after the one day of intensive activity. Rather, by providing a fairly complete report of the activity of the day, not only are the recommendations themselves available, but also the context in which they were developed. In this way it is expected that others can use the recommendations as they are or modify them as they wish.

The two-year curriculum developed here was seen as integrated in the sense that topics in discrete mathematics and those in continuous mathematics were to be woven together throughout the two year period. Integrated was not taken to mean that the entire two year course was an integrated whole. In fact, it was important that each individual quarter or semester of material be a unit ending at an appropriate place, which might well be an appropriate stopping point for one group of students.

Additionally, it was assumed that all students had completed the equivalent of four years of the traditional high school mathematics prior to beginning the courses; and that they had learned to program a computer prior to the courses or were taking a programming course concurrently with the first semester (quarter).

The two-year course was designed in so far as possible so that it could coordinate with the curriculum of its major users: engineering, computer science, biological sciences, physical sciences, and social science. Further, the recommendations were to be of value for a small liberal arts college that could only offer one program or for a larger institution with a variety of possible tracks.

Students often change majors and an effort was made to design a two year program that would minimally penalize a student who changed major.

Faculty teaching the courses of this integrated program were assumed not only to have a knowledge of programming, but to be aware of the possible usage of existing and emerging technology related to hand calculators, computers, and computing devices with the facility to perform symbolic manipulations. The necessity for a core faculty group within a department with these skills, who act as advocates for the course was noted, as was the need for appropriate means to train other faculty. Many members of the mathematics faculty will be faced with two hurdles. First they will have to learn new material not a part of their own training which they have not taught before; and second they will have to become familiar with the use of the computer and its appropriate use as a teaching tool.

Method of development of the course recommendations

During one day of intensive activity after establishing the above as a basis for course development, the workshop participants followed a process described below which led to the development of course recommendations. It should be carefully noted that the recommendations are based on the prior experience of the participants, the papers and the discussions of earlier portions of the conference, and one day of intensive discussion. They are not seen as a final clear mandate, but rather as a basis for further discussion. They are "first approximations" to a possible two year integrated curriculum, and it is hoped that they will lead to discussion, trial courses, and further recommendations. The procedure which led to these recommendations was:

I. Five week modules were seen as the basic unit of information. A list of twenty modules that might be appropriately included in the first two years of collegiate mathematics was developed. Six other modules were discussed but not included. These were identified either as third year courses (which might be taken earlier) or as alternatives for the first two years.

II. A diagram showing prerequisites for each of the modules was developed. (See p.) It was noted that prerequisites are of two types: a preferred order that would affect language and style of presentation, or those needed because of their content.

III. Nine disciplinary areas which might require (or recommend) two years of mathematics for certain majors in the discipline were noted. These were biological science, business and economics, computer science, industrial engineering, mechanical engineering, electrical engineering, mathematics, physics, and social science. For each of these nine disciplines, those modules from the set of twenty that were seen as most appropriate for these disciplines were listed. Comparison of these lists led to a consolidation into five disciplinary areas: mathematics, computer science, physics/engineering, biological science, social science (including business and economics). All of the engineering disciplines were lumped together. However, industrial engineering with a slightly different modification might have been placed with the mathematics or computer science list.

IV. An appropriate sequencing of modules was developed for each of the five consolidated areas. These sequences took careful note of both preferred and necessary prerequisites and of the timing of presentation of material to maximize its use to the discipline.

The sequencing of the modules led to groupings of the modules into courses of three modules each.

V. The semester courses developed in IV tended to differ from discipline to discipline. (See diagram on p.) A final stage was the development of a set of semester courses (three modules each) that were modified from those in IV in order to unify the semester offerings from discipline to discipline. These were used to develop a four semester sequence for each disciplinary area. It should be noted that the development at this stage placed further constraints on content and on the preferred order of presentation. Thus, although this set of semester courses met the guidelines established, i.e. they can lead to a unified course of study for a liberal arts college, or to several tracks for a university, and to a reasonable set of options without too much penalty for students who change major, there was some loss in the suitability of the sequences for each of the related majors, some loss of best placement of preferred prerequisites, and some further compression of material.

The material presented below follows the above five steps. The reader of this article wishing to modify these courses to fit her/his own perceptions or the curricular needs of a particular program might most easily begin with I. using an adaptation of our process.

I. A list of twenty five-week modules appropriate to the first two years of college mathematics

Each participant in the workshop developed her/his list. The union contained twenty-six modules. The module development in Bushaw's paper was seen as a motivating force for this discussion. After discussion, six modules were not included (see below) in the final recommendations. It should be noted that no attempt was made to carefully delineate content (except for G and I), but there were discussions, without full consensus of opinion, as to whether and at what cost the modules could be developed.

Note that the list omitted the letter O. Further, P--Algorithms, Q--A Review of Algebra and Geometry to Include Conics and Higher Algebra (such as in Hall and Knight) [2], and U--Introduction to Algebraic Structures (Systems, Groups, Semigroups), were not included in the final list, but seen as material for a third year. The Modules V--Modeling, W--Cultural Liberal Arts, and X--Mixed Applications, although seen as potentially appropriate for the first two years, were not made a part of the final recommendations.

The following brief descriptions were given for G and I in order to answer the question of prerequisites:

G. Matrices, systems of linear equations, Gaussian elimination, inverses, determinants--brief treatment of the iterative definition by cofactors, eigenvalues, diagonalization

I. R^n, linear independence, basis, dimension, subspace, vector space, projections, least squares, inner products, orthogonalization, linear transformation

This delineation and further discussion led to a decision that I should be a third year course and it was therefore dropped from all of the recommendations with the understanding that some material might be included in G and K, the latter divided into K_1 and K_2.

A list of all the modules

A. Sets, functions, relations, recursion, induction, sequences, limits

B. Derivative and its applications (mention optimization), exponential and trigonometric functions

C. Integration

D. Infinite series, improper integrals (Bushaw #5)

E. Partial derivatives

F. Multiple integration (include use of the computer and symbolic manipulation)

G. Concrete linear algebra

H. Vector calculus

I. Abstract linear algebra (Note: Omitted from final recommendations.)

J. Mathematical logic

K. Continuum equations--differential equations, integral equations

(Note: Split into two full modules after discussion.)

K_1. Linear independence and differential equations

K_2. Differential equations and integral equations

L. Probability

M_1. Statistics (five-week overview)

Note: Each of the modules M_2, M_3, and M_4 are full modules.

M_2. Data taking, organization, description, sampling

M_3. Exploratory data analysis

M_4. Statistical inference

N. Algorithms, induction, recursion, sample applications to calculus, concept of complexity

P. Algorithms

Q. Review of algebra, geometry (conics, higher algebra)

R. Graphs

S. Combinatorics (with discrete probability?)

T. Linear programming and inequalities

(Note: Split into two modules after discussion.)

T_1. Linear programming and linear inequalities

T_2. Combinatorial optimization

U. Introduction to algebraic structures (systems, groups, semigroups)

V. Modeling (existence, uniqueness, complexity)

W. Cultural liberal arts, applications to daily life (discrete and continuous)

X. Mixed applications

II. A diagram showing the prerequisites for each module

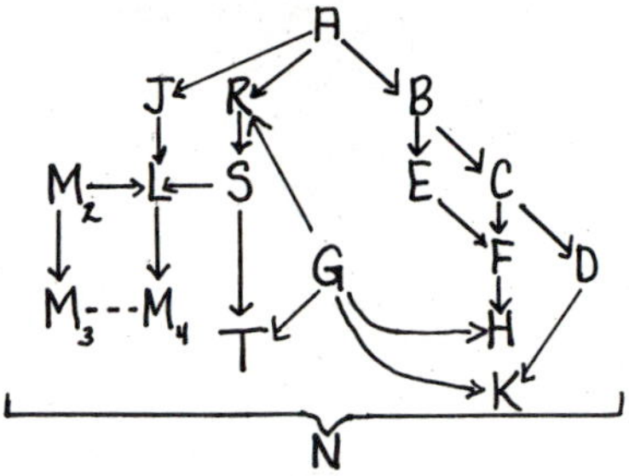

KEY:

A. Relations
B. Differential calculus
C. Integral calculus

D. Infinite series
E. Partial Differentiation
F. Multiple Integrals

G. Linear Algebra
R. Graphs
S. Combinatorics

H. Vector Calculus
K. Differential/ Integral equations

J. Logic
M_2. Data
L. Probability

M_3. Data analysis
M_4. Statistical inferences
T. Optimization

Some consideration was given to teaching integration before differentiation. This order was seen by some as preferable since the idea of sum (area) was seen as simpler than that of derivative, and because the computation of approximations to integrals leads to an appropriate introduction to a relationship between discrete and continuous mathematics. The books by Donald Richmond [3], Robert L. Wilson [4], and Tom M. Apostol [5], were cited and the advantages for a course which combines discrete and continuous mathematics were noted. If this approach was used then the order would be ACB.

III. The modules seen as appropriate for each of the nine disciplinary areas

BIO. SCI.	ABC	-	-	GI	-	J	-	L	M_2UM_4	-	N	RS	T
SOC. SCI.	ABC	-	-	GI	-	J	-	L	M_2UM_4	M_3	N	RS	T
PHYSICS/ MECH. ENG./ ELEC. ENG.	ABC	D	EF	GI	H	J	K	L	M_2UM_4	-	N	RS	T
INDUS. ENG.	ABC	D	E	GI	H	J	-	L	M_2UM_4	M_3	N	RS	T
COMP. SCI.	ABC	-	-	GI	-	J	-	L	M_2UM_4	-	N	RS	T
MATH.	ABC	-	EF	GI	-	J	-	L	M_2UM_4	-	N	RS	T
BUS. ECON.	ABC	-	E	GI	-	J	-	L	M_2UM_4	M_3	N	RS	T

IV. Sequences of modules recognizing prerequisites, grouped into semesters, for mathematics, computer science, physics and engineering, and biological sciences

Semester	First	Second	Third	Fourth
Mathematics	ABC	M_2GR	JSL/DEF	M_4IT
Computer Science	AGJ	RST_2	BCM_2	LT_1I
Physics/Engineer.	ABC	GM_2D	JEF/RST	K_1K_2H
Biological Science	AGM_2	BCJ	RSL	$M_4T_1T_2$

Social Science was not treated at this point. I was dropped from all disciplines and necessary material from I therefore had to be included in G and K. K was divided into two full modules, K_1 and K_2. In the mathematics and physics/engineering sequences, two concurrent courses are scheduled for the first half of the second year.

All the participants strongly preferred the courses developed here to those developed later in V. The five-week modules made for a flexibility that allowed maximum adaptation for each field or purpose. Further, A was used as the initial course in all sequences and was therefore seen as the place to set a tone and an approach for the entire sequence. A as an appropriate beginning, consisting of a joining of the continuous and discrete approaches, would be highly desirable.

V. A set of semester (three module) courses and their preferred order as a sequence for each of the five principle disciplinary areas.

The module approach above provides a most appropriate choice for each discipline. If courses could be presented as five week mini-courses then there would be greater flexibility to meet the needs of each discipline. Although textbook publishers might develop some books on a module basis, teaching courses in the modular system has several drawbacks. It could be an administrative nightmare. Further, it would make transfer from one major to another difficult, and though giving several possible patterns to a liberal arts institution, it does not state one simple pattern. Therefore six semester (three module) courses were devised based on a clustering of courses, illustrated in the following diagram.

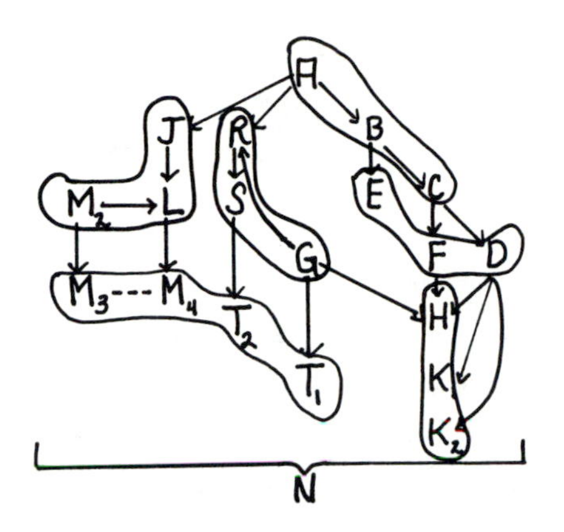

KEY:

A. Relations
B. Differential Calculus
C. Integral Calculus

D. Infinite Series
E. Partial Differentiation
F. Multiple Integrals

G. Linear Algebra
R. Graphs
S. Combinatorics

H. Vector Calculus

K_1. Linear independence/ differential equations
K_2. Differential equations/ Integral equations

J. Logic
M_2. Data
L. Probability

M_3. Data analysis
M_4. Statistical inferences
T_1. Linear programming/ linear inequalities
T_2. Combinatorial optimization

The sequences for the five areas would be:

Semester	First	Second	Third	Fourth	Later semesters		
Computer Science	G'RS	A'BC	JM_2L	M_3M_4T	N	DEF	HK_1K_2
Mathematics	ABC	GRS	JM_2L/ DEF	M_3M_4T	N	HK	
Physics/Engineering	ABC	GRS	JM_2L/ DEF	HK	N	M_3M_4T	
Biology	ABC	GRS	JM_2L	M_3M_4T	N	DEF	HK_1K_2
Social Science	G'RS	JM_2L	A'BC	M_3M_4T	N	DEF	HK_1K_2

The computer science and the social science sequences would begin with discrete mathematics. Since A (relations, sets, functions, etc.) would not be the initial module, some modification might have to be made in G, hence the notations A' and G'. Students in mathematics, computer science, physics, engineering and biology would have covered the same material by the end of the first year.

The integration of discrete and continuous mathematics in these recommendations is not within the semester (as was the case in certain of the modular courses), but by the inclusion of a discrete semester in the first year and at the beginning of the second year. However, in each presentation of each semester all material should tie together and build the interrelationship (marriage) of discrete and continuous mathematics.

Main Themes--Course N

During discussion of the papers presented in the early part of the conference, several themes were established as important to the integrated approach to discrete and continuous mathematics. These themes were algorithms, induction-recursion, linearity, optimization, "conjugation," function-relation, estimation, geometry and abstraction. Each course of the sequence will have topics that develop certain of these themes. The style of presentation in every course should emphasize the themes and this emphasis should provide a viewpoint and approach for both discrete and continuous sections of the course.

In addition to the recurrent play of the themes throughout the earlier courses, Course N was seen as a summary course, a look back at the themes and the topics that display these themes. There might be a "Course N" for each discipline. Algorithms, induction, and recursion as well as other themes might be discussed abstractly then illustrated from material presented earlier.

Conclusion

Had time permitted, a set of quarter system courses constituting sequences for each discipline would have been developed. These might allow one to put back into the courses some of the flexible approach seen in the modular sequence.

The workshop group discussed a number of topics related to an integrated course and reached a consensus on the following. First, the integrated nature of the two year sequence must be established by the approach throughout. Second, if the modular approach can be encouraged to the extent that textbook publishers will have authors develop modular units that can be included in a variety of semester or year-length books, then experimentation and course development can take place broadly. Third, it was noted that a variety of "new" course concepts that were discussed had actually been tried before (e.g., calculus with integration presented first; the course GRS:

linear algebra, graphs, combinatorics).

Fourth, repeatedly the group noted the tentative nature of their recommendations. They felt strongly about many of the ideas at the time, but had no opportunity for reflection. Further, many of the details of the course content were not discussed. Finally, there was genuine enthusiasm on the part of all concerned for the development of a two year sequence integrating discrete and continuous mathematics.

REFERENCES

1. Committee on the Undergraduate Program in Mathematics, Recommendations for a General Mathematical Sciences Program (Washington D.C.: Mathematical Association of America, 1981).

2. H. S. Hall and S. R. Knight, Introduction to Algebra, (New York: Macmillan, 1928).

3. Donald Everett Richmond, Calculus, (New York: McGraw-Hill, 1950).

4. Robert Lee Wilson, Much Ado About Calculus, (New York: Springer-Verlag, 1979).

5. Tom M. Apostol, Calculus, (Waltham, Massachusetts-Blaisdell, 1967).

Report of the Workshop on IMPLEMENTATION

Workshop members: Richard Anderson, William Lucas, Lynn Steen, Alan Tucker, Gail Young, Stephen Maurer (chairman)

Our workshop had an easier job than the others; we didn't have to deal with specific mathematical topics and come to a consensus about them. Nonetheless, outlining a "plan of attack" for the future is clearly important, and we hope we have come up with enough specific suggestions to be useful.

We began with the following question: Are we, the movement to experiment with change in the first two years of the undergraduate mathematics curriculum, far enough along to start writing texts and looking for schools to undertake courses? Among us, we had two answers to this question -- Yes and No! No, because we don't yet know enough about what will be acceptable to other disciplines; at this conference we did have representatives from other disciplines to give us their reactions, but they were a very small and highly non-random sample. Also, if the wrong sort of book is the first to come out, negative reaction to it could kill any hope of further developments. Yes, because we already have such a diversity of opinions that it is hard to reconcile them; if we ask for too many more, the issues may become so hopelessly muddled that it will be impossible to find a clear path. Sometimes it's just best to forge ahead, as Kemeny, Snell and Thompson did, for, if a good course and text appear, all the opinion givers (who after all don't really have a clear idea of what they want) will decide that <u>that</u> is it!

Given this difference right at the start, there was obviously only one plan we could agree to: attack on all fronts simultaneously! So that is what we propose, with the following Principle and following

Caution. Principle: we cannot force an institution to try a new curriculum; all we can do is keep the issue highly visible and encourage tempted institutions to try a change. Caution: the more radical a change we propose, the less likely institutions are to want to try it. Gradualism seems to work best, so a program with two distinct year courses, one on continuous mathematics and one on discrete, is more likely to be tried than a program with one integrated 2-year course. That way a school can begin by simply adding the discrete course, rather than changing everything. In fact, during most of our discussions, we quite consciously thought only of how to implement this simpler sort of change, because many of us thought a more radical change was simply not going to happen. Consequently, because the other discipline most interested by far in having us serve them with a discrete course is computer science, most of our thoughts about interacting with other disciplines were deliberately limited to interacting with computer science.

We were, of course, aware that the separate course scenario is not really just a matter of adding a course. The one-year continuous mathematics course is supposed to condense 3 semesters of calculus. Whether both changes would or should be made simultaneously, or whether a school would first introduce the discrete course and later try condensing the calculus, is not something it occurred to us to discuss. But we did explicitly consider whether a school was likely to convert its whole curriculum to the new approach, or whether it would run the new in parallel with the old. We believed only the latter was likely -- putting all the eggs in one basket is too risky -- but there was disagreement on this when we reported to the conference as a whole (see below). Once we realized we were thinking about parallel tracks in any event, there no longer seemed to be such a difference in likelihood between the separate and the integrated approaches. Also, the workshop charged with creating an integrated curriculum ended up presenting its proposal in terms of semester modules, so we don't really have a 2-year take-it-or-leave-it package to "sell".

In short, then, we see the job ahead as consisting of spreading the word, collecting information, and obtaining schools, text materials and funds to try initial programs. Below we proceed to each of these items in turn.

Spreading the Word. We need to keep the proposals for change constantly before the mathematical community, and before the other disciplines we service. In particular, we ought to keep in mind that many of the people who are interested in the uses and effects of computers in education are not active in the mathematical societies, but rather in various computer and computer education societies. In any event, a brief report about this conference ought to appear in several places: the MAA Focus, the AMS Notices, the SIAM Newsletter, and the Newsletter of ACM's SIGCSE (the Special Interest Group for Computer Science Education of the Association for Computing Machinery). Certain individuals ought to be informed too, for instance, James Stasheff of the University of North Carolina (Chairman of the Subcommittee on Education of the AMS Science Policy Committee), Ronald Wenger of the University of Delaware (Director of CIMSE, the National Consortium on Uses of Computers In Mathematical Sciences Education), and Gerald Engel of Christopher Newport College (a leader in national computer education activities). Other groups and individuals need to be identified and informed.

In addition, discussions ought to be scheduled at MAA meetings. A specific proposal for the next national meeting appears under Collecting Information below. Here we point out that the regional meetings of the MAA are much better for group discussions: there is a different and wider clientele, the numbers in attendance at sessions are more conducive to discussion, and there are many such meetings instead of one. The MAA regional officers should be encouraged to put such sessions on their programs.

We ought also to be sending out emissaries to meetings of other groups. For instance, the next annual meeting of the National Computer Education Conference is in June in Towson MD. Also, in October James Fey of the University of Maryland is holding a conference on the high school mathematics curriculum, in which major changes will be discussed. There are surely many more such conferences at which the changes we propose should be presented.

Collecting Information. As stated at the outset, we need to know more about what mathematics is important in the eyes of the disciplines we serve, especially computer science. We also need to know which mathematicians and which institutions are strongly interested in trying new courses, or may actually be trying them already, without any fanfare about it. The latter information is easier to obtain; we have already placed an announcement in Focus asking for faculty who are already teaching algorithmic discrete mathematics courses at the freshman-sophomore level to "report in" to MAA headquarters. Obtaining the former information will take more work. First, we suggest that a mathematics delegation be sent to a few selected institutions of various sorts (a liberal arts college, a large state university, an institute of technology, a school with a strong business orientation, etc.) to talk specifically to the computer scientists (and perhaps others) at these places. Next, we recommended that a joint MAA-ACM committee be set up to discuss these curricular issues. There has obviously been need for close communication between these two societies for some time, and yet not much has happened because there has never been a formal mechanism. Next, we suggest that a set of mathematicians be designated to study the question of what mathematics is really behind undergraduate computer science. Computer science has its own jargon; although there is no doubt at least a partial bijection between CS terms and terms in mathematics, lack of knowledge of the bijection has kept most mathematicians from really understanding what computer scientists are talking about, and thus from knowing what mathematics really ought to be taught for computer science. Mathematicians currently being retrained to be computer scientists would be ideal choices to work on this project.

Clearly, there needs to be a focal point for all such curriculum outreach activities. The MAA is obviously the right organization, and there is an obvious entity within the MAA -- the Committee on the Undergraduate Program in Mathematics (CUPM) The CUPM has been very busy with several other projects, but fortunately several of these have recently been completed. With a new chairman about to come in (after the long and very effective leadership of Donald Bushaw, who attended this conference), it is a good time for CUPM to commence a major new project. Both Anderson (the MAA President) and Bushaw were quite receptive to this idea. The usual procedure for CUPM is not to work on an issue as a whole, but to set up a panel. We recommend this. The panel might be, or might include, a joint committee with ACM, as suggested above. The appropriate committee of the ACM to hook up with is probably the Curriculum Committee for Computer Science.

As a final information gathering activity, we propose that one of the contributed papers sessions at the coming MAA winter meeting in Denver be devoted to reports on courses already being given, or planned to be given soon, of the sort we are talking about. Indeed, this subject might become a recurrent topic for such MAA sessions.

<u>Finding Schools at Which to Implement</u>. First of all, "schools" is not really the right focus. We ascribe to the "great man" theory. What we need to find is a single person at each of several institutions who is dedicated to this experiment and highly energetic; he or she will carry the ball. Furthermore, we don't think it will be hard to

find such people. Already, almost every mathematics department has at least a small minority who are waiting to be encouraged in the directions this conference has proposed.

What sort of institution will provide the most fertile ground? We feel the small liberal arts colleges will. First, faculty there regard educational issues as part of their professional responsibilities, and thus such schools are way ahead of the research-oriented universities in thinking about curricular developments. Second, such small schools generally have not created separate computer science departments; the mathematics faculty are teaching what computer science is taught and so are already familiar with the issues that CS raises for mathematics education. Third, in order to attract students, who might otherwise go to larger and better known institutions, such schools like to have distinctive programs with which to advertise themselves. Fourth, being small, the problems of coordination with faculty in other departments (discussed at length for large universities in Barrett's paper) don't much exist.

Our workshop recognized one problem with spearheading the proposed curricular changes at small undergraduate schools. In as much as faculty there are not so actively involved in research, they are often behind in understanding what changes in mathematical perspective are taking place and why. Another, perhaps more serious concern about small schools was raised at the time of our oral report: being small, did such schools have the resources to run the traditional and new courses in parallel? It was certainly felt they wouldn't have to invent extra teaching loads; even the smallest schools currently run 2 or more sections of the basic underclass courses. Whether the burdens of running more types of courses with fewer sections each would deter such schools is not something we all had the same intuition about. Anthony Ralston, who has visited many such schools in the last year, felt strongly that they would want to give one sequence only, and that many such schools were very interested in <u>replacing</u> the traditional sequence with the new one. Being very interested in this, and actually doing it, are two different things, but in general the conferees all reached the same optimistic conclusion that a sufficient number of colleges would be willing to give the new courses a try.

Two dangers with parallel programs should be pointed out. First, if students have a choice between the two programs, the new program may never get a proper test. If the new curriculum is ever to become <u>the</u> curriculum, then the crucial test is whether physical scientists and engineers will come to think of it as appropriate for them. If the traditional program is now given in parallel, science and engineering students may not try it, and hence never get used to it. Also, because in the new curriculum students can (perhaps should!) put off calculus for a year while they take discrete math, the students who choose to take the new curriculum may be the weaker students. Therefore, when faculty in various disciplines later compare the "graduates" of the two sequences, they may conclude erroneously that the new sequence is less good at fostering mathematical maturity. One way to overcome these problems is to make the discrete mathematics part of the new curriculum a prerequisite for later courses that able math, science and engineering students take, e.g., linear algebra (which should build on the matrix algebra in the discrete course), differential equations (which should build on the linear algebra), probability and statistics. We strongly recommend that schools take this step.

Second, and perhaps more important as a danger, is that having parallel tracks may force students to make career choices too early. On the other hand, if the new track really is good preparation for most scientific majors, then there there isn't much of a danger here.

Returning to types of schools, it would obviously be good if some large universities also experimented, and we felt it would not be hard to find some that would -- at the instigation of their computer science departments. CS departments desperately need to have mathematics share their teaching load; ACM in its current curriculum report says that the basic mathematics of computer science should be taught by mathematics departments. Furthermore, CS departments are looking for ways to increase the professionalism of their students, who are often more interested in making money with a computing career than they are in the intellectual aspects of the discipline. Many CS departments would love to have math departments offer a course or two of relevant mathematics which they could require CS students to take, and which would therefore increase the students' professional level. The fraction of faculty in university mathematics departments sympathetic to the computer scientist's view may be smaller than in colleges, but again, if one or two faculty are interested, a program can be started.

Texts. Usually, in order for publishers to be interested in new texts, there has to be a pre-existing university clientele, that is, a course already existing with a fairly large national enrollment. However, publishers seem hot to provide new types of discrete mathematics texts. Even if no discrete course is presently being given of quite the sort an author proposes, the publishers know that a latent market is there, and they are eager to get to it first. So getting publishing support is no problem. Also, several people are writing or thinking of writing discrete texts; perhaps the course outlines provided by the other two workshops will be of help to them.

Unfortunately, we did not discuss the prospects for texts for the new one-year continuous mathematics course -- whether there is publisher interest and whether there are writers-in-waiting. The need for such books must be made known, especially to publishers.

Another issue is how to get publishers interested in providing text material for the modularized integrated sequence proposed by the

Barrett workshop. The word "modules" is anathema to publishers; they have been burned by attempts to market modules in the past, that is, to market pamphlets covering one or two lectures worth of supplementary material. But this is not what that workshop means by modules. Modules for them come in five week minicourses. We need to make this distinction clear to publishers. It was suggested that we not use the word module, but rather find some other term.

In any event, we cannot expect much finished text material to be available as the first experimental courses start up. The teachers of these first courses may have to provide their own written materials. In other words, they may also become the text writers.

Funding it all. As just stated, support for production of finished texts should not be much of a problem. What is needed are funds to support information gathering, and development of trial courses and initial written materials at individual schools. First we discuss where this support might come from and then how much is needed.

As for NSF, the traditional primary source, there is no hope of any help from there in the next year or so. There is a glimmer of hope that a significant part of the Education Directorate funding of past years may be restored by Congress after that, but we shouldn't count on it.

The only major private foundation which has funded research and education in mathematics and basic science during the last 10 years is the Sloan Foundation. Our aims are very much up Sloan's alley, and part of our needs might well be funded as part of, or in conjunction with, Sloan's major new program "The New Liberal Arts". This is a 15 million dollar, 5 year program to place mathematical and technological literacy in the core of the liberal arts, by encouraging faculty in all disciplines to use mathematical analyses in their teaching wherever appropriate. However, with so much of Sloan's resources now committed to this general education program, it remains to be seen whether Sloan will want to commit more than token funds to a purely mathematical project.

There are of course other foundations to try, foundations which have shown at least some interest in science education matters. Those mentioned by conferees were the Lilly Endowment, the Exxon Foundation, the MacArthur Foundation, and the Johnson Foundation. The last named has so much money and is still so new that it hasn't yet figured out what to do with it all! Also mentioned were the IBM and Xerox Corporations. For instance, Xerox has been interested in publishing modules. Finally, it was asked if computer companies themselves might not help. However, the prospects for that do not seem good; everybody is asking them for help, and many of them are relatively tight financially at the moment. Except for giving away a few micros, they probably won't be very interested.

How much funds are needed? For the work we propose CUPM should do, 5 or 10 thousand. That's not much as requests go, and as it is for seminal activities with broad visibility, we felt there is a very good chance funding can be obtained.

For developing a new discrete mathematics course, we reasoned as follows. To provide sufficient incentive and released time for people to do the job well, there should be a stipend for the summer before the course is first given, one course released time during the year it is first offered, and a stipend for the following summer to make revisions. We were thinking in terms of one person (the great man theory), so this sounds like $25,000. However, we thought it best that the money be given to the department or school, and let them decide if it should go to one person or more. Finally, it would be good to have 5 to 10 schools try such experiments simultaneously. True, when one thinks in terms of trying an integrated sequence instead of a mere course, $25,000 is perhaps low, but in any event, we thought that a good national experiment could get under way for a quarter million.

One way to proceed would be to convince some foundation(s) to entertain proposals for $25,000 or so, and then let the foundation(s) choose among the applicants. Another way would be for CUPM to obtain all the available money as a block grant, and then have it decide how to allot the money among schools.

We emphasize that such funding would only be for start-up. Great individuals may get things going, but if discrete mathematics is to be a co-equal with calculus, then before long everybody must teach it. As has been pointed out elsewhere, the algorithmic discrete syllabus we have in mind will require a considerable change of perspective on the part of most mathematicians. We cannot expect to bring about such a change by paying to retrain everybody. Rather, there has to be a snowballing effect. As a new curriculum gets more popular, and as computing becomes more pervasive, more and more faculty will want to absorb the new perspective, and will do so in part by self-study and in part by osmosis.

Moral support. Several workshop members thought it important to get strong statements of support for curricular change from the MAA and the AMS. When mathematics departments go to deans to get additonal internal funding for their curricular experiments, and surely they will have to do this, they may not get much sympathy if the deans find that the mathematics professional organizations have not strongly urged such changes.

As far as getting an AMS statement is concerned, it was pointed out that the AMS has not (by tradition) been involved with undergraduate education. However, one member felt it was very important to get a statement from the AMS because in some sense it is viewed as the most prestigious mathematics society. Furthermore, he pointed out, in the future there won't <u>be</u> many people going into university mathematics research, the AMS bailiwick, unless something is done to change our educational program. Thus it would be in the direct interest of the AMS to give support. Whether the AMS will view things this way remains to be seen, but we have already noted that it is taking a more active interest in education through the Stasheff committee.

As for the MAA, President Anderson observed that it would be premature for it to advocate the general adoption of this new curriculum. Its appropriate role now is the spreading, gathering and evaluation of information he had already promised to pursue. Furthermore, the MAA does not make curriculum recommendations except through the CUPM. If in time the CUPM decides to advocate general adoption, fine.

Anderson also pointed out that the National Academy of Science is planning to increase its presence in educational issues, so perhaps some strong support will be forthcoming from there in time.

Finally, in answer to the unsympathetic dean, a member suggested it would be enough to point out that the changes the mathematics department proposes would result in a stronger offering in computer science; deans even more than students want their schools to be au courant in this!

<u>Other matters</u>. There were two topics which our workshop was charged to consider but about which our members felt they need say little. The first was: what objections might be raised to the proposed curricular changes, and how might they be answered? We felt that Ralston has already cataloged these objections, and good answers to them, in his various articles. Furthermore, we felt that if the new curriculum is indeed run in parallel with the traditional one during the initial stages, this will deflect most criticism. Finally, we

propose always taking the positive approach when there are objections. Pointing out, as in the previous paragraph, that we will now be better at meeting the needs of computer-oriented students, is such an approach.

The second charge was: how can the effectiveness of a new curriculum be evaluated? Our workshop did not wish to get bogged down in disputations on the value of formal research in education, and fortunately, neither do those faculty who doubt the wisdom of changing from the traditional calculus curriculum. That is, so long as a few years from now our colleagues in mathematics and other disciplines feel _intuitively_ that the students who have been through the new curriculum are as well trained (or better trained!) than traditional students, we will have met the evaluation test. One caution though: we must make our colleagues (especially in physical science and engineering) aware that "trained as well" does not mean "trained the same". These graduates of the new courses will be trained differently, and there will be topics which are now sometimes assumed in later courses which these students will not know, or not know so deeply. But if their general grasp of clear thinking, and the variety of techniques they can bring to bear on problems is at least as good as for traditional students, then we ought to agree that they are trained as well.

One final remark, raised during our oral report: we have not paid any attention during this conference to junior colleges. Many universities receive a substantial number of their students as transfers from junior colleges. If we don't get junior college faculty involved in our curricular changes, down the road there will be a clash at the interface between these two types of institutions.

INDEX